Plastics
vs. Corrosives

175 YEARS OF PUBLISHING
1807 1982

SPE MONOGRAPHS

Injection Molding

Irvin I. Rubin

Nylon Plastics

Melvin I. Kohan

Introduction to Polymer Science and Technology: An SPE Textbook

Edited by Herman S. Kaufman and Joseph J. Falcetta

Principles of Polymer Processing

Zehev Tadmor and Costas G. Gogos

Coloring of Plastics

Edited by Thomas G. Webber

Analysis and Performance of Fiber Composites

Bhagwan D. Agarwal and Lawrence J. Broutman

The Technology of Plasticizers

J. Kern Sears and Joseph R. Darby

Fundamental Principles of Polymeric Materials

Stephen L. Rosen

Plastics Polymer Science and Technology

Edited by M. D. Baijal

Plastics vs. Corrosives

Raymond B. Seymour

Plastics vs. Corrosives

RAYMOND B. SEYMOUR

CMA Distinguished Professor
Department of Polymer Science
University of Southern Mississippi
Hattiesburg, Mississippi

Professor of Chemistry Emeritus
University of Houston
Houston, Texas

A WILEY-INTERSCIENCE PUBLICATION

JOHN WILEY & SONS

New York • Chichester • Brisbane • Toronto • Singapore

Library of Congress Cataloging in Publication Data:

Seymour, Raymond Benedict, 1912–
Plastics vs. corrosives.

(SPE monographs, ISSN 0195-4288)
"A Wiley-Interscience publication."
Includes bibliographical references and index.
1. Plastics—Corrosion. 2. Plastic coating.
I. Title. II. Title: Plastics versus corrosives.
III. Series.

TA455.P5S454 620.1′92323 81-21996
ISBN 0-471-08182-5 AACR2

Printed in the United States of America

10 9 8 7 6 5 4 3 2 1

Series Preface

The Society of Plastics Engineers is dedicated to the promotion of scientific and engineering knowledge of plastics and to the initiation and continuation of educational programs for the plastics industry. Publications, both books and periodicals, are major means of promoting this technical knowledge and of providing educational materials.

New books, such as this volume, have been sponsored by the SPE for many years. These books are commissioned by the Society's Technical Volumes Committee and, most importantly, the final manuscripts are reviewed by the Committee to ensure accuracy of technical content. Members of this Committee are selected for outstanding technical competence and include prominent engineers, scientists, and educators.

In addition, the Society publishes *Plastics Engineering, Polymer Engineering and Science* (*PE & S*), *Journal of Vinyl Technology, Polymer Composites,* proceedings of its Annual, National, and Regional Technical Conferences (ANTEC, NATEC, RETEC), and other selected publications. Additional information can be obtained from the Society of Plastics Engineers, 14 Fairfield Drive, Brookfield Center, Connecticut 06805.

Robert D. Forger
Executive Director
Society of Plastics Engineers

Preface

In a report on the use of plastics in corrosive environments some 30 years ago, it was stated that the annual cost of corrosion in the United States rivals the interest on the national debt. Unfortunately, neither the magnitude of the interest on the national debt nor the expense of corrosion has subsided. According to a recent report by the National Bureau of Standards, the annual loss due to the corrosion of metals in the United States is now over $70,000,000,000.

Of course, these costs would be much higher if chemical resistant plastics were not available to combat corrosion and to replace corrodable metal structures. It is now recognized that industrial corrosion cannot be accepted as a normal operation either from an economic or environmental viewpoint.

Current technical information on plastics available three decades ago was supplied in reports to the American Chemical Society* and the Society of Plastics Engineers† as well as in a comprehensive book.‡ The volume and variety of available plastics and reports on their use in corrosive environments has grown stupendously since those inceptive reports were published. This new book has been written in an attempt to update the states of the art, technology and science in the important field of chemical resistant plastics.

It is hoped that readers of this book will find it to be a useful tool in the solution of corrosion problems.

RAYMOND B. SEYMOUR

Hattiesburg, MS
March 1982

* R. B. Seymour, *Modern Plastics* **27**(11), 19 (1951).

† R. B. Seymour, *SPE Journal* **10**(14), (1954).

‡ R. B. Seymour and R. H. Steiner, *Plastics for Corrosion Resistant Applications*, Reinhold Publishing Co., New York, 1951.

Acknowledgment

The author appreciates the suggestions made by Otto H. Fenner, who reviewed the manuscript before publication.

RAYMOND B. SEYMOUR

Contents

Plastics vs. Corrosives

CHAPTER 1

Introduction

The ancients used bitumens for waterproofing their boats and shelters. After the discovery of corrosives, it was observed that the bitumens used as caulking materials were not only resistant to water but also to vinegar, niter and "fuller's scope." However, the use of these organic materials as containers for either water or corrosives was limited. The Chinese and Egyptians used ceramics and alchemists used glass as containers for their crude chemical processing, and in spite of their inherent brittleness, these products are still used as materials of construction today.[1]

That natural rubber (*Hevea braziliensis*) was also resistant to water was recognized by the Indians in Central and South America who made waterproof boots and bottles from the product obtained from *Cau-uchu* or "weeping wood." In his abortive attempt to waterproof 150 mailbags for the United States government in the 1830's, Charles Goodyear discovered that rubber and not cotton was resistant to nitric acid fumes. Goodyear reduced the tackiness of the rubber coated mailbags and life preservers by his "acid gas" process but the fabric in these articles, which was not resistant to nitric acid, deteriorated after a relatively short period of time.

The cellulose trinitrate produced by Schönbein in 1846 by heating cellulose with nitric and sulfuric acids proved resistant to additional amounts of these acids. Celluloid, which Hyatt produced in 1868 by the plasticization of cellulose nitrate using camphor, was also resistant to many corrosives.

Menard demonstrated that cellulose nitrate was soluble in a mixture of ethanol and ethyl ether and this solution, called collodion, was used as a protective coating in the mid-nineteenth century. Prior to the 1960's, the metal surfaces of most automobiles and other outdoor equipment were protected by the application of a pigmented solution of cellulose nitrate.

Bakelite, the world's first truly synthetic plastic, was used in the early 1900's as an acid resistant molded plastic.[2] Laminated plastic made by molding phenolic-resin-impregnated paper or cotton fabric has been used to a limited extent in corrosive environments for over a half a century.

Saureschutz Gesellschaft in Berlin produced chemical resistant equipment in the 1920's by molding a mixture of asbestos and phenolic resin.[3] This product, under the trade names of Faolite and Haveg, is still available today. A corrosion resistant mortar, called by the trade name Asplit, was also produced in the 1930's by mixing a resole phenolic resin with a silica filler containing benzene sulfonyl chloride which served as a latent source of acid.[4] Other chemical resistant mortars have been produced by using furan, epoxy, and polyester resins in place of the phenolic resole and by using ground coke in place of the silica filler.

Hard rubber, the first man-made plastic, has been used for molded acid resistant containers for over a century. Vulcanized natural rubber has also been used as a corrosion resistant lining for steel vessels for over 50 years.

Although shellac, bitumens, cellulose nitrate lacquers and phenolic resins[5] have been used as protective coatings, the significant advances in this field were the result of the introduction of chlorinated rubber coatings under the trade name of Tornesit, in Germany in the 1920's and coatings of the copolymer of vinyl chloride and vinyl acetate in the United States in the 1930's. These are also still in use.

An annual review of plastics as materials of construction was published by the American Chemical Society in *Industrial and Engineering Chemistry* from 1947 to 1971.[6] These annual reviews have been published in *Australian Plastics and Rubber* and *Popular Plastics* (Bombay) after the publication of *Industrial Engineering Chemistry* was discontinued.[7,8]

Considerable history has been recorded but much more will be written as new test methods and more case history data on applications of plastics in corrosion resistant environments are developed. Plastics play a unique role in this field and their use will increase as more corrosion engineers become acquainted with plastics and more plastics engineers become acquainted with problems in the corrosion field.

REFERENCES

1. R. F. Eisenberg, R. R. Kraybill, and R. B. Seymour, Chapter 1, "Selection of Materials for the Construction of Equipment," *Laboratory Engineering and Manipulation,* 3rd ed. (E. S. Perry and A. Weissenberger, eds.), John Wiley and Sons, New York, 1979.
2. L. H. Baekeland, U.S. Patent 957,137 (1910).
3. E. Elbel, German Patent 533,286 (1927).
4. K. Kietz, K. Frank, F. Privinsky, E. Thiel, German Patent 595,014 (1934).
5. H. Klas, U.S. Patent 1,932,691 (1933).
6. G. M. Kline and R. B. Seymour, *Ind. Eng. Chem.* **42,** 2004 (1950).
7. R. B. Seymour, *Aus. Plast. Rubber* **26**(12), 9 (1975).
8. R. B. Seymour, *Pop. Plastics* **23**(1), (1978).

CHAPTER 2

Fundamentals of Plastics

As shown by the following skeletal formulas (2.1), all organic compounds contain carbon atoms that are usually covalently bonded to other carbon atoms or to oxygen, nitrogen, phosphorus, or sulfur.

C—C	C—OH	C—NH_2	C—O—C	(2.1)
Ethane	Methanol	Methylamine	Dimethyl ether	

Of course, inasmuch as carbon is a tetravalent atom, hydrogen or other monovalent atoms such as chlorine will also be covalently bonded to the carbon atom as shown by the more complete structural formulas (2.2). However, for reasons of simplicity, skeletal formulas will be used in this book.

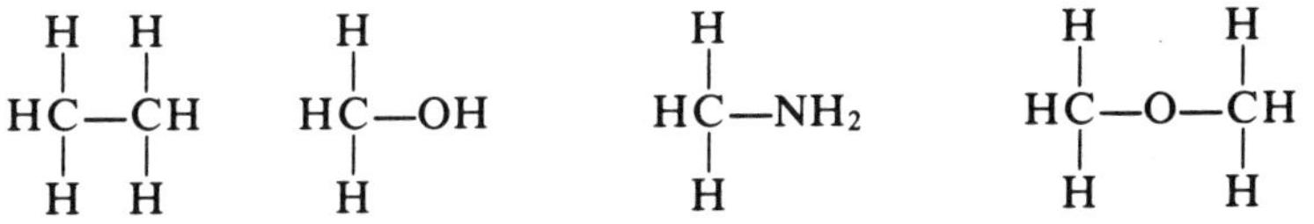

(2.2)

Ethane   Methanol   Methylamine   Dimethyl ether

The average distance between the carbon atoms in ethane is 0.154 nm and the bond angles are approximately 109°. The average bond energy or the energy required to break a carbon–carbon bond and a carbon–hydrogen bond in ethane is 80 and 99 kcal, respectively. In addition to these strong primary valence bonds, there are also weaker intermolecular forces ranging from less than 1 kcal for London or dispersion forces such as those present in ethane to as much as 10 kcal for strong hydrogen bonds such as those present in methanol.

The simplest organic polymer is linear polyethylene (HDPE) which, as shown by the skeletal formula of a segment of the polymer (2.3), is simply an extension of the formula shown for ethane (2.1).

—C—C—C—C—C—C—C—C— (2.3)

HDPE is said to be linear because this polymer consists primarily of a continuous carbon–carbon chain.

As shown in the following structural formula (2.4), it is customary to show the repeat unit in brackets with a subscript number to indicate the average number of repeat units or mers. The letter *n*, which is also the degree of polymerization ($\overline{DP}$), may represent a lower limit of about 50 for a useful structural polymer and a value of 1000 or many thousands for the upper limit.

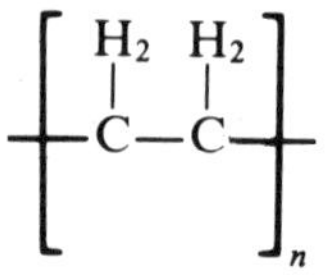

(2.4)

The preceding statements made for simple molecules (which may be considered as model compounds) also apply to polymers. Thus, the atoms in the principal chain are covalently bonded. The average bond energies of the carbon–carbon and carbon–hydrogen bonds, the average bond distance for the carbon–carbon atoms, and the bond angles are similar to those cited for small organic molecules.

The hydrogen atoms on one polyethylene chain are attracted to the atoms in another polyethylene chain by weak London or dispersion forces. Inasmuch as the forces are additive, they are greater for a polymer than for a small molecule of comparable structure, that is, a model compound.

Whether a polymer is classified as a elastomer, plastic, or fiber is dependent on the relative strength of the intermolecular forces and the geometry of the molecule. The intermolecular forces in an elastomer are usually weak London forces and there are no stiffening groups in the chain. There are usually substituents in place of the hydrogen atoms in elastomers (i.e., pendant groups) on the principal chain or back bone. This is illustrated by the skeletal formula for a segment of polyisoprene (natural rubber), a typical elastomer (2.5):

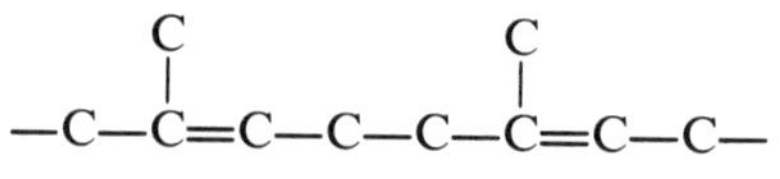

(2.5)

An elastomer in its unstretched state is amorphous or noncrystalline. In contrast, fiber molecules, which have good geometrical symmetry, are usually crystalline. The repeat unit of nylon-66 (2.6) contains amide stiffening groups ($-\overset{\overset{\displaystyle O}{\|}}{C}-\overset{\overset{\displaystyle H}{|}}{N}-$) and will have strong intermolecular hydrogen bonds between the hydrogen atoms of the amide group on one chain and the oxygen of an amide group on another chain and the oxygen of an amide group on yet another chain.

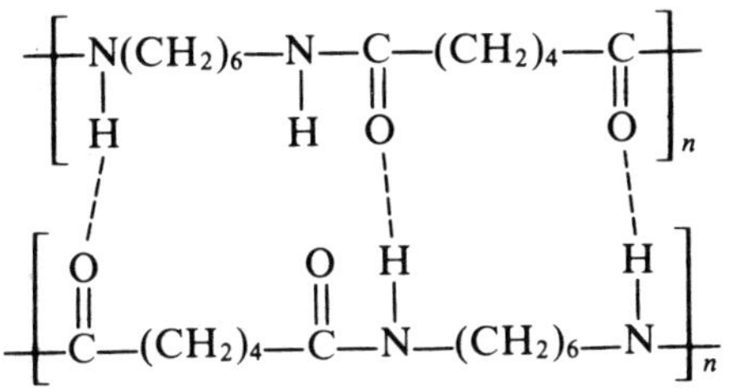

(2.6)

Most plastics are amorphous polymers with properties between those of elastomers and fibers. However, fibrous nylon-66 can be used as a molded plastic and some polyolefin plastics can be used as elastomers.

Regardless of whether a polymer is classified as an elastomer, plastic, or fiber, its chain length must be long enough to permit entanglement. Thus, although paraffin wax has the same formula as polyethylene, it has little structural strength because of its relatively short chain length.

As shown by the skeletal formula for low density polyethylene (LDPE) (2.7), some polymers have branches that are randomly placed chain extensions of the repeat unit. A functional group on the repeat unit, such as the methyl group of polyiosprene (2.5), is called a pendant group and not a branch.

Skeletal formula for a segment of LDPE:

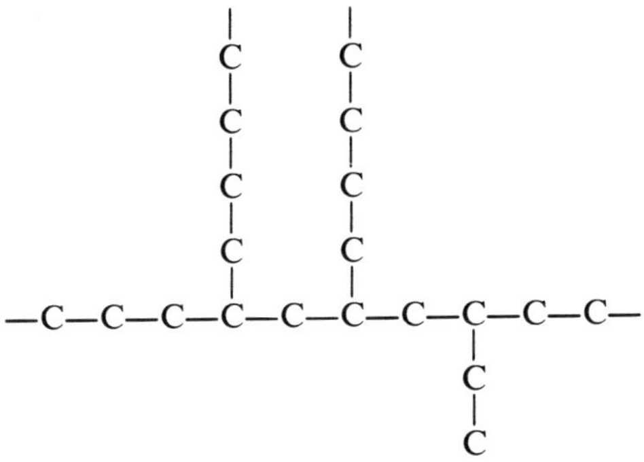

(2.7)

With the exception of polymers like cellulose, which have strong intermolecular hydrogen bonds, most linear and branched chain polymers are softened by heat and hence are called thermoplastics. The temperature at which segmental motion of the chain in an amorphous polymer begins is called the glass transition temperature (T_g) or second order transition temperature. Hard plastics become rubber-like when heated above T_g. The melting point (T_m) or first order transition temperature of a polymer is always higher than T_g.

When the polymer chains are connected to each other by primary valence bonds, the crosslinked or network polymer is insoluble in solvents and not softened by heat without decomposition. These crosslinked polymers (2.8) are called thermosetting plastics.

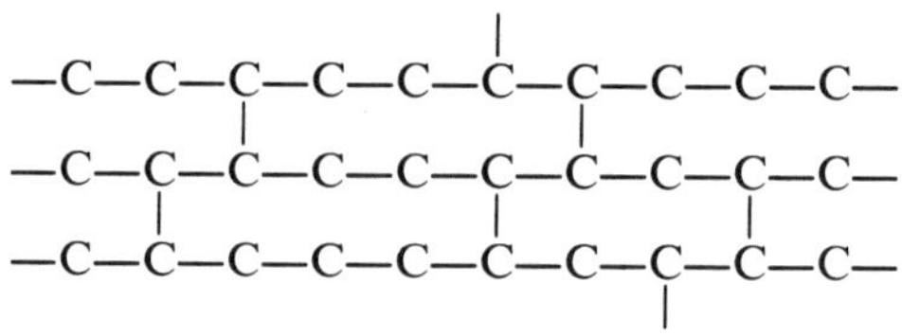

(2.8)

This, of course, represents a simulated structure of a segment of a crosslinked polymer.

Nitrogen atoms are present in the principal chain of polyamides (2.6) such as nylons and proteins. Likewise, oxygen atoms are present in the principal polymer chain of polyesters such as polybutylene terephthalate (PBT) and in acetals such as polyoxymethylene (POM). Some polymers such as polysiloxanes (silicones) do not have any carbon atoms in the principal chain (2.9). However, most of the principles discussed for organic polymers apply to these inorganic polymers.

A segment of the principal chain of a silicone polymer:

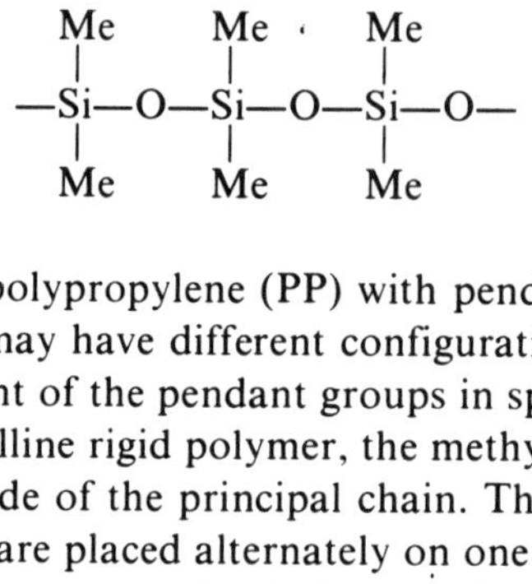

(2.9)

Polymers such as polypropylene (PP) with pendant groups on alternate carbon atoms (2.10) may have different configurations or tacticity depending on the arrangement of the pendant groups in space. In the isotactic PP, the commercial crystalline rigid polymer, the methyl pendant groups are all placed on the same side of the principal chain. The pendant groups in the syndiotactic polymer are placed alternately on one side and then the other. These pendant groups occur randomly in space in atactic PP which is a soft amorphous viscous liquid.

Isotactic PP:

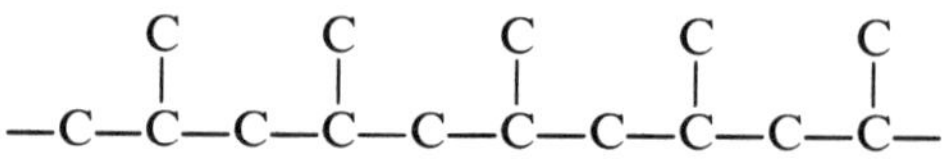

Syndiotactic PP:

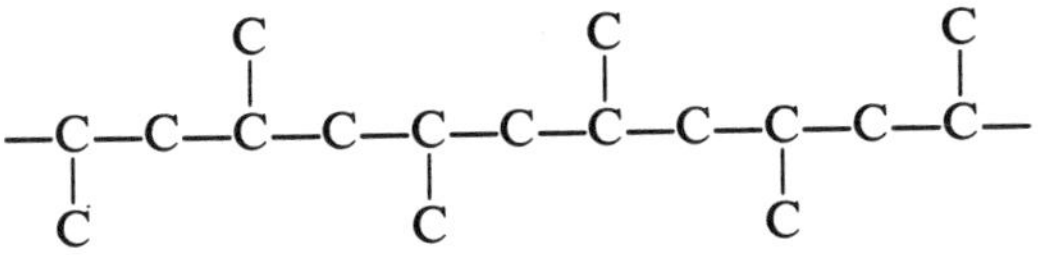

(2.10)

Atactic PP:

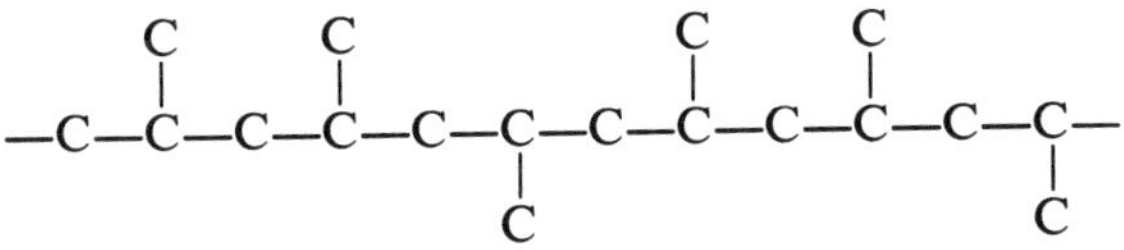

The polyolefin plastics consist of the polyethylenes (LDPE and HDPE) and polypropylene (PP). They are produced at an annual rate of 3.5, 2, and 1.5 million tons, respectively, and are the most widely used plastics. Because of the absence of nonhydrocarbon pendant groups, they are also resistant to many corrosive environments.

The segments of other large volume thermoplastics, such as polyvinyl chloride (PVC), polystyrene (PS), polymethyl methacrylate (PMMA), and polyurethane (PU) are shown in (2.11). The annual production of PVC, styrene polymers, and acrylic polymers in the U.S.A. is 2.8, 2.4, and 0.15 million tons, respectively. Almost 900,000 tons of polyurethane are produced annually in the United States but most of this polymer is used as flexible (608,000 tons) and rigid foams (258,000 tons).

Segments of major nonolefinic thermoplastics:

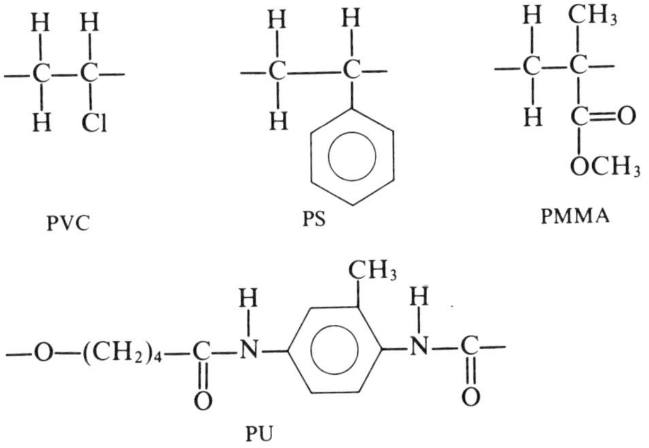

(2.11)

The annual production in thousand tons of cellulose acetate (CA), polyphenylene oxide (PPO), nylon (plastics), polyacetal (POM), polycarbonate (PC), thermoplastic polyesters (PET, PBT), and thermoplastic elastomers (TPE, styrene-butadiene-styrene copolymers, SBS, Kraton), is 225, 68, 140, 48, 109, 408, and 139, respectively. The formulas for segments of these thermoplastics are shown below.

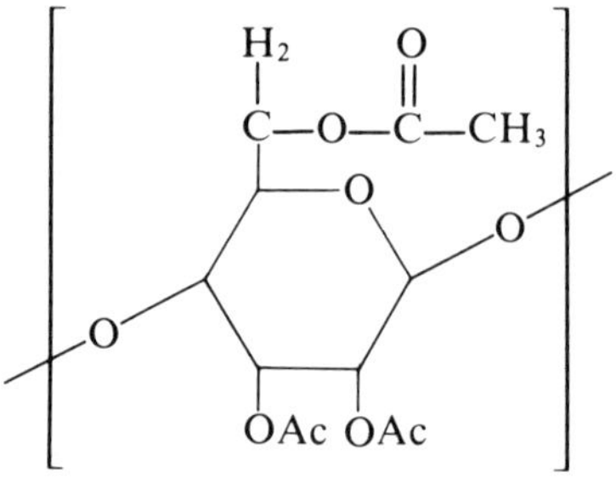

CA

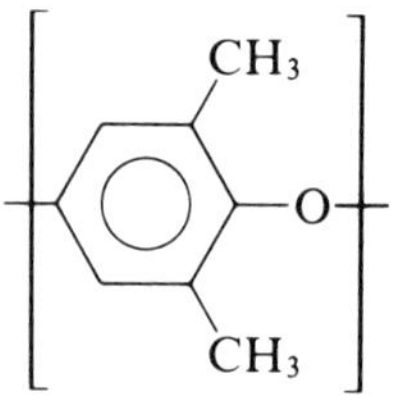

PPO

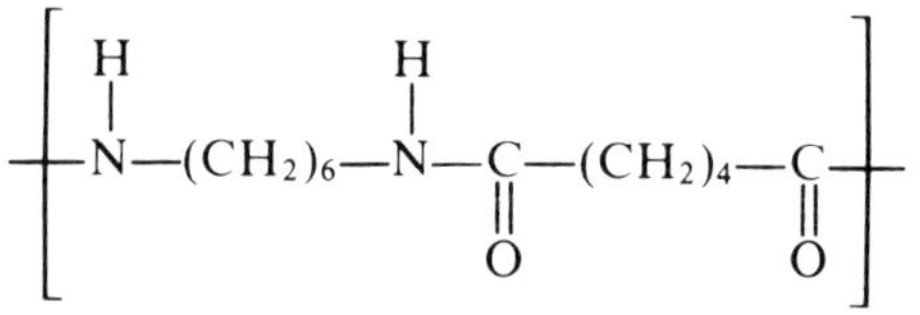

Nylon-66

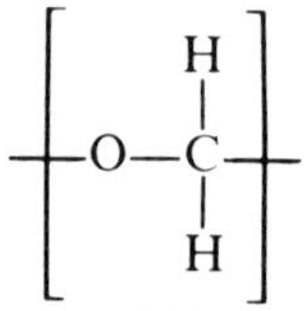

POM

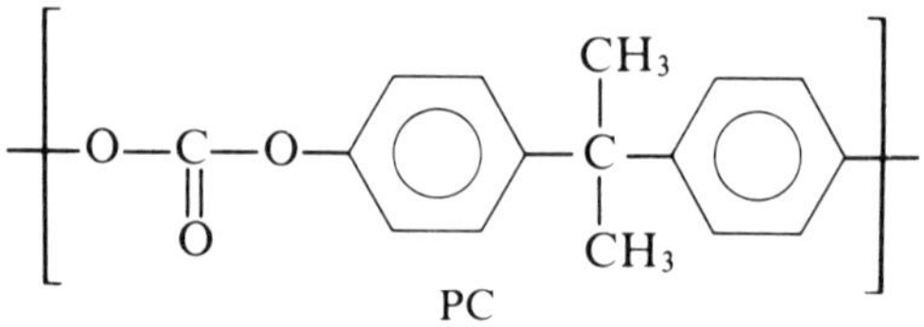

PC

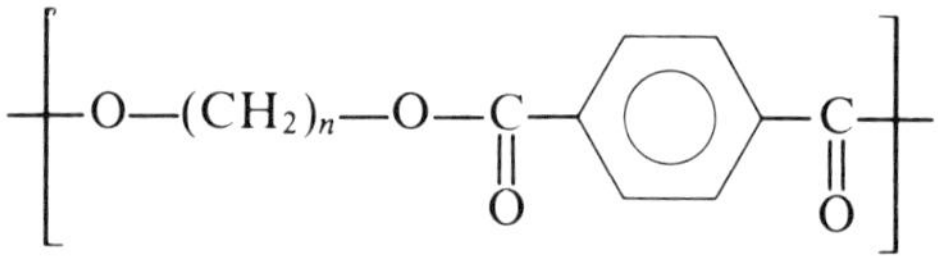

PET ($n = 2$), PBT ($n = 4$)

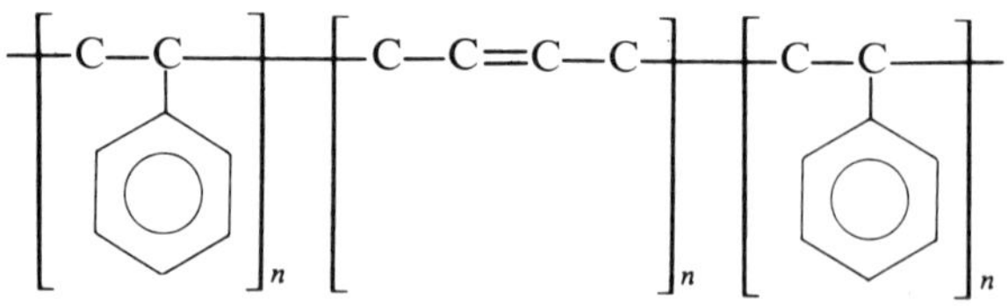

(2.12)

SBS copolymer (TPE)

The phenylene () shown in the formula for PU, PPO, PC, PET, and PBT, is a stiffening group, whereas the methylene group ($-CH_2-$) shown in PET, PBT, and many other polymers is a flexibilizing group. The flexibilizing action is related to the number of methylene groups between the stiffening groups. Thus, polybutylene terephthalate (PBT), which has four methylene groups, is more flexible than polyethylene terephthalate (PET), which has two methylene groups.

Polyolefins PVC and PMMA are called aliphatic polymers, whereas those with phenyl () groups such as PS, PU, PPO, PC, PET, and PBT, are called aromatic polymers.

SBS (2.12) is produced by the polymerization of more than one monomer. Since both styrene and butadiene are present as repeat units in the chain, this is called a copolymer. The repeat units in a copolymer may be randomly or alternately spaced, or they may be in multiples. The structure of the copolymer of vinyl chloride (87) and vinyl acetate (13, Vinylite) is a random copolymer and that of styrene and acrylonitrile (2.13) is an alternating copolymer.

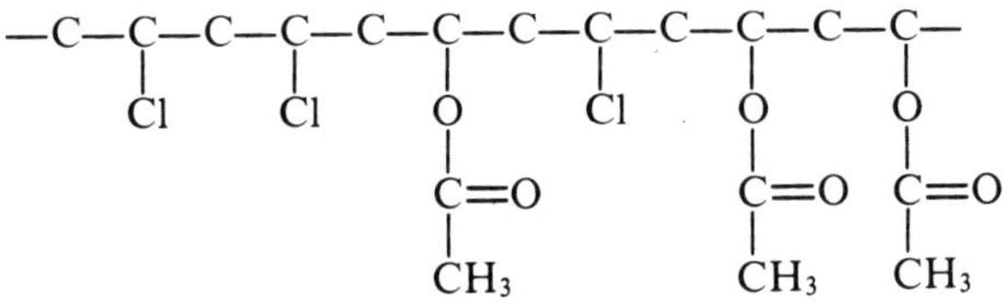

Vinyl chloride–vinyl acetate copolymer

(2.13)

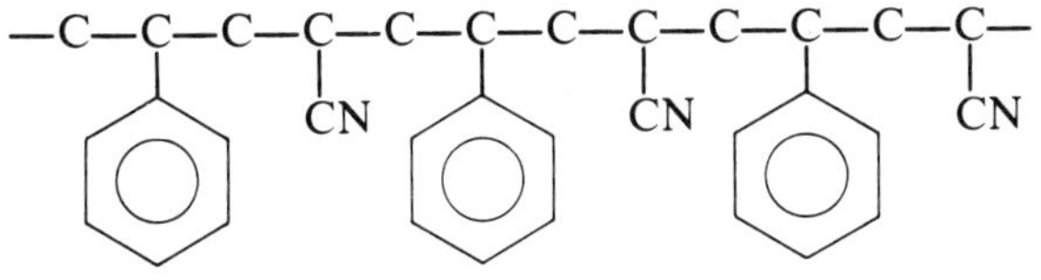

Styrene–acrylonitrile copolymer

The SBS copolymer (2.12) is a block copolymer in which sequences of styrene repeat units are followed by a sequence of butadiene repeat units and followed again by a sequence of styrene units. Another copolymer, in which the building units or mers exist in multiples, is in graft copolymers, such as ABS copolymers. The chain extensions (2.14) made up of a sequence of another monomer are located as branches on the chain. The term terpolymer may be used to indicate the presence of three monomers in the copolymer.

Acrylonitrile–butadiene–styrene (ABS) terpolymer, a graft copolymer:

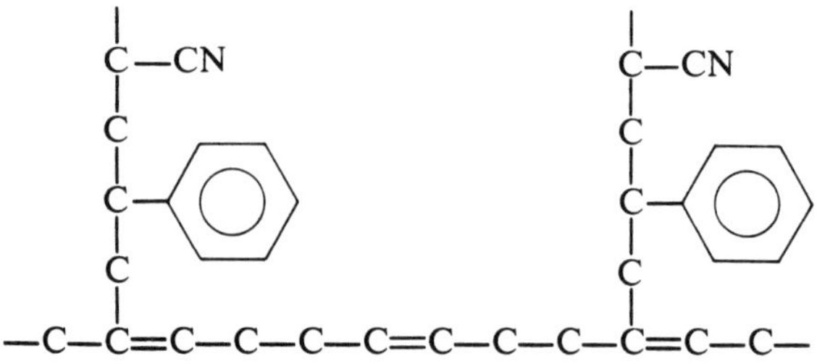

(2.14)

ABS

Phenolics, urea and melamine resins, unsaturated polyesters, alkyds, and epoxies are produced at annual rates of 704, 582, 560, 225, and 155 thousand tons, respectively. The formulas for segments of these thermosets (2.15) follow.

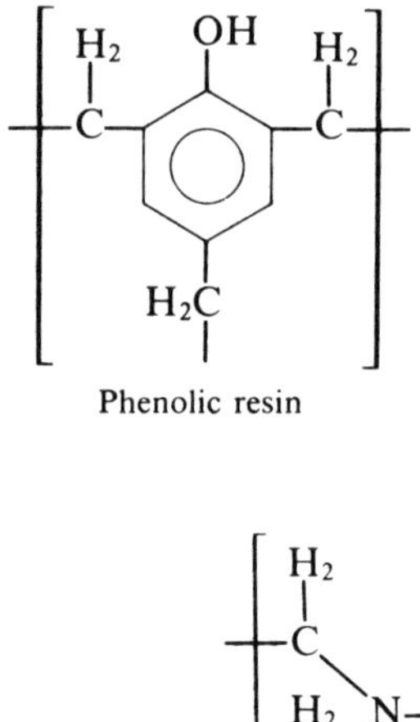

Phenolic resin

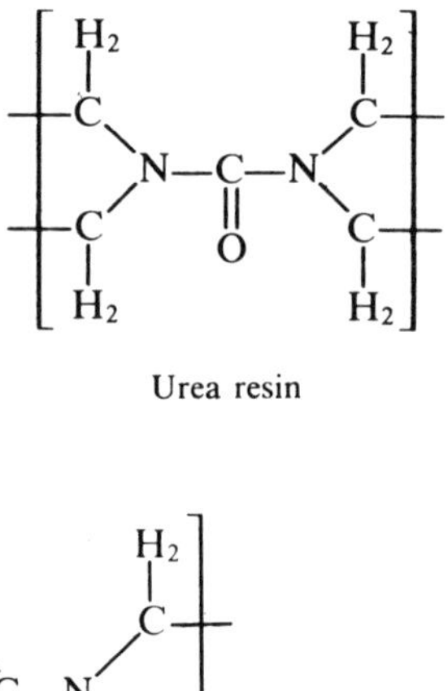

Urea resin

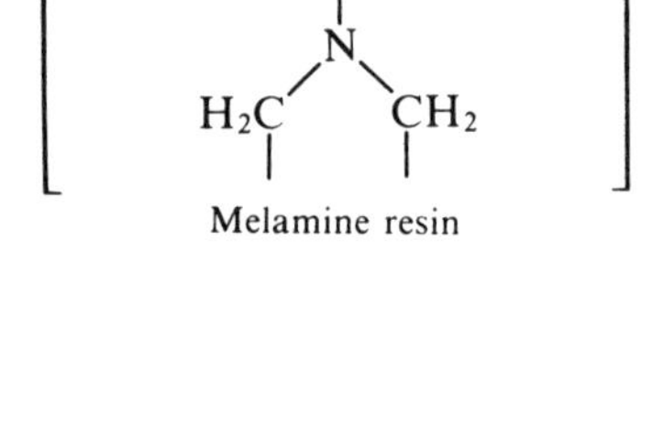

Melamine resin

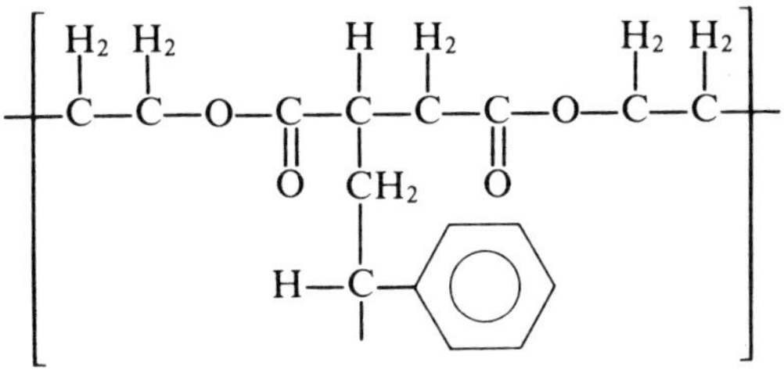

Unsaturated polyester resin

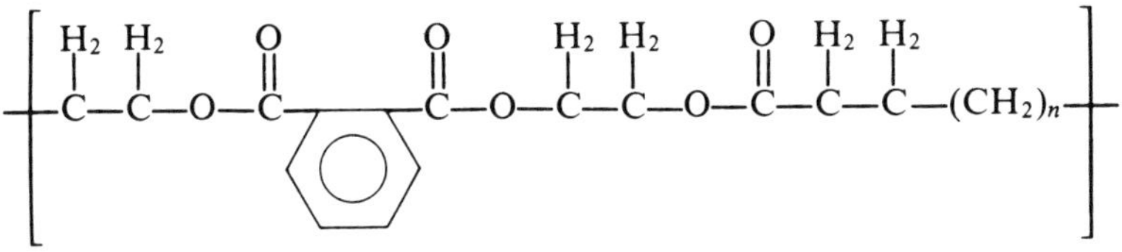

Alkyd resin

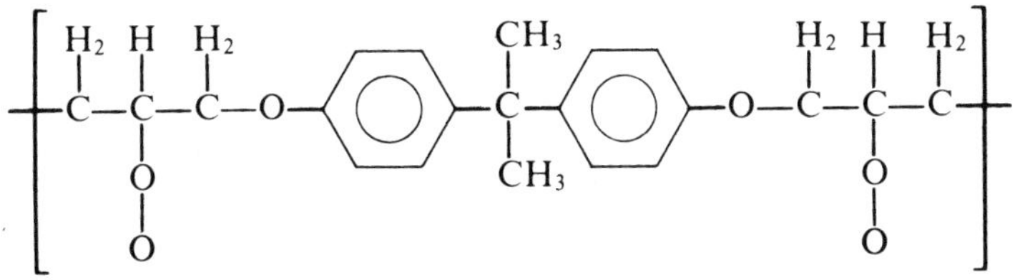

Epoxy resin

(2.15)

The formula for an epoxy resin (2.15) is for one cured at elevated temperatures by a reaction with an acid anhydride such as phthalic anhydride. However, most epoxy plastics used in corrosion resistant applications are cured by polyamines which react with the terminal oxirane (epoxy) groups and do not react with the hydroxyl groups.

The unsaturated polyester resins may be made from bisphenol-A instead of ethylene glycol. The bisphenol-A may also be esterified with acrylic acid instead of maleic acid. Segments of these modified unsaturated polyester resins (2.16) are shown below.

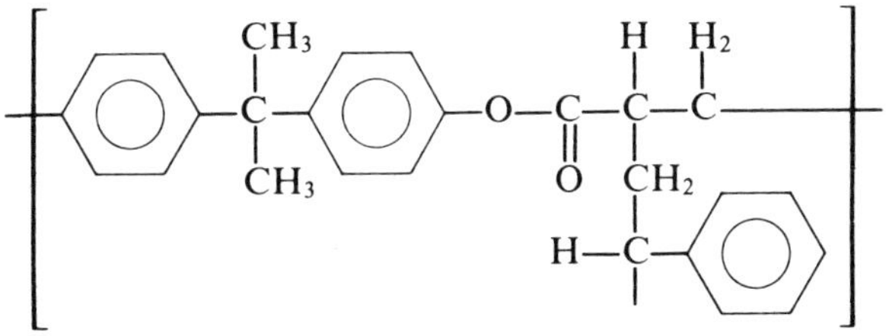

Bisphenol-A polyester

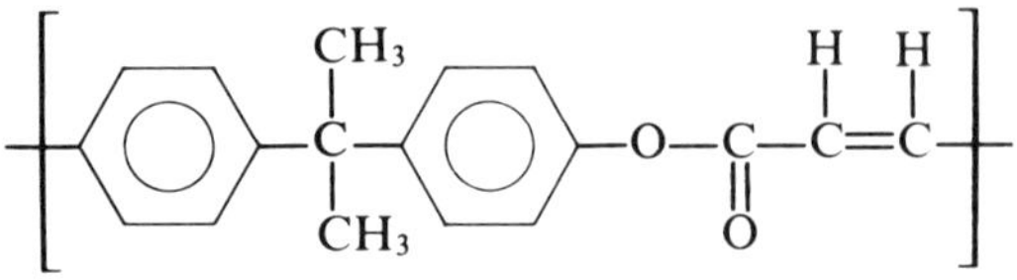

(2.16)

Vinyl ester resin

It is important to note that, as a general rule, the reaction of polymers is similar to that of their small molecule counterparts, that is, model compounds. Nobel laureate, Paul Flory, has postulated an equal resistivity principle. Although this rule may be used as a guide, the reaction of corrosives with solid polymers is dependent on the accessibility of the reactive groups to the corrosive. Such reactions are hindered by low absorption or penetration of the corrosive, by crystallinity, and by neighboring groups (steric hindrance). These reactions are also enhanced by temperatures above T_g and by solvation effects.

Plasticizers such as dioctyl phthalate, which are added to PVC to improve flexibility, also enhance the reactivity of corrosives both with the plasticizer and the polymer. In contrast, additives such as fillers and reinforcements, which raise the T_g and are bonded to the polymer, may increase the resistance to corrosives.

Changes in structure, such as the substitution of a fluorine atom for a hydrogen or chlorine atom, will increase the resistance of polymers to chemical attack. The average bond energy of the carbon–fluorine bond is 102 kcal. Thus, polytetrafluoroethylene (Teflon) is resistant to the action of most corrosives. However, since this polymer is not soluble in solvents and is difficult to mold, several other fluorocarbon polymers have been commercialized.

The following formulas are for segments of these chemically resistant polymers (2.17):

Polytetrafluoroethylene (PTFE, Teflon)

Polychlorotrifluoroethylene (PCFE, Kel-F)

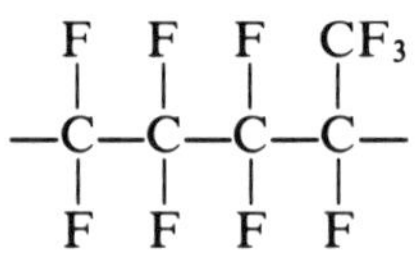

Tetrafluoroethylene–hexafluoropropylene copolymer (FEP)

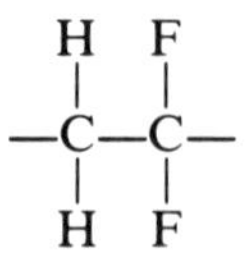

Polyvinylidene fluoride (PVDF, Kynar)

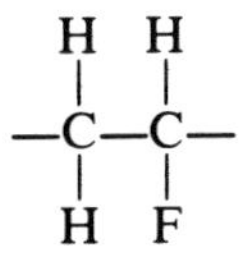

Polyvinyl fluoride (PVF, Tedlar)

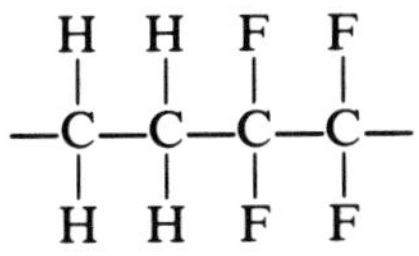

Ethylene–tetrafluoroethylene copolymer (ETFE, Tefzel)

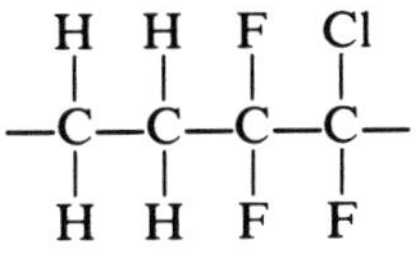

(2.17)

Ethylene–chlorotrifluoroethylene copolymer (ECTFE, Halar)

Additional information on fundamentals of plastics may be found in standard polymer textbooks, some of which are listed at the end of this chapter.

REFERENCES

1. F. W. Billmeyer, *Textbook of Polymer Science*, 2nd ed., John Wiley & Sons, New York, 1971.
2. J. Brandrup and E. H. Immergut, *Polymer Handbook*, John Wiley & Sons, New York, 1975.
3. J. M. G. Cowie, *Polymers: Chemistry and Physics of Modern Materials*, Intext Educational Publishers, 1973.
4. J. P. Flory, *Principles of Polymer Chemistry*, Cornell University Press, Ithaca, NY, 1973.
5. H. S. Kauffman and J. J. Falcette, *Introduction to Polymer Science and Technology*, SPE Monograph, John Wiley & Sons, New York, 1977.
6. R. N. Lenz, *Organic Chemistry of Synthetic High Polymer*, Interscience Publishers, New York, 1967.
7. G. Odian, *Principles of Polymerization*, McGraw-Hill, New York, 1970.
8. F. Rodriquez, *Principles of Polymer Systems*, McGraw-Hill, New York, 1970.
9. R. B. Seymour, *Introduction to Polymer Chemistry*, McGraw-Hill, New York, 1971.
10. R. B. Seymour, *Introduction to Polymer Chemistry*, American Chemical Society, Audio Course, Washington, DC, 1980.
11. R. B. Seymour, *Modern Plastics Technology*, Reston Publishing Co., Reston, VA, 1975.
12. R. B. Seymour and C. E. Carraher, *Polymer Chemistry: An Introduction*, Marcel Dekker, New York, 1981.
13. M. P. Stevens, *Polymer Chemistry*, Addison-Wesley Publishing Company, Reading, MA, 1975.
14. B. Vollmert, *Polymer Chemistry*, Springer-Verlag, Berlin, 1973.
15. H. L. Williams, *Polymer Engineering*, Elsevier Publishing Co., New York, 1975.

CHAPTER 3

Physical Properties of Polymers

The chemistry of pure polymers was discussed in Chapter 2, yet it must be emphasized that although there are a few exceptions such as some fibers, films, and sheets, structural polymers are usually composites. Additives such as heat and light stabilizers, fillers, plasticizers, pigments, coupling agents, lubricants, fungicides, bacteriocides, and antistats are often present. Therefore, the plastic under test will usually be one that is designed to perform better than the pure polymer in actual use in hostile environments.

Since materials of construction are purchased by the kilogram and used by the cubic meter, specific gravity or relative weight per unit volume is an important property. Although the specific gravity of filled plastics is usually greater than unity, that of most unfilled plastics is approximately equal to one.

In contrast, the specific gravity of most other materials of construction (metals, ceramics, or concrete) is greater than one. Thus when reported on a volume or area basis, strength values may be biased in favor of nonplastics. However, the strength properties of plastics are more competitive when the comparative strength data are on a weight basis.

Hardness is also an important property and with the exception of wood, the hardness values of plastics are usually inferior to those of other materials of construction. Thermal properties of plastics are also usually inferior to those of metals and ceramics but thermosets and some filled thermoplastics are useful at moderately high temperatures such as 150°C.

Most nonpolar plastics such as HDPE are poor conductors of heat and electricity. The dielectric constant (the ratio of the capacitance of a parallel plate condenser of the plastic to the capacitance in a vacuum or more commonly in air) is an important property when the plastic is used as an electrical insulator. However, it is important to note that metal-filled plastic composites and special polymers such as polysulfur nitride and polyacetylene are conductors of electricity.[1,2]

Unlike metals and like glass, many unfilled plastics are transparent or translucent. Although this may not appear to be an important property in corrosive environments, it is advantageous to use transparent plastics for glazing, lenses, gauges, and even piping. Most plastics are less brittle than glass. The index of refraction (n) is a characteristic value that may be used to identify polymers.[3]

Because of the emphasis on noise abatement, acoustic properties of polymers are also important.[4] Properly selected plastics serve as absorbers of sound and as damping devices.[5] Of course, high frequency ultrasonic waves may destroy plastic structures.

When the plastics are moving parts such as in pumps, or are used as pipes to convey materials, friction and wear (i.e., the science of tribology) are important considerations.[6] As the coefficient of friction is related to the critical surface tension of a plastic, PTFE, which has a critical surface tension of 18 dyn/cm, has an extremely low value for static coefficient of friction (0.04).

It has been postulated that the surface of PTFE consists of rolled up chains with exposed fluorine atoms that provide surface lubricity. PTFE serves as a lubricant when used as an incompatible finely divided filler in plastic composites. The low friction characteristics of nylon-66 and polyurethane are believed to be the result of a monomolecular layer of water which is hydrogen bonded to the plastic surface.[7]

The elastic component is dominant in plastics below T_g but the viscous component is also present. Therefore, plastics are said to be viscoelastic materials. The elastic components may be described by Hooke's law and the viscous component by Newton's law.

When an amorphous plastic is subjected to a moderate stress (S), the carbon–carbon bonds in the polymer are stretched or compressed and the normal tetrahedral angles of 109° are distorted. According to the law for an ideal elastic solid developed by Hooke in 1876, S is proportional to the resulting elastic deformation or strain (γ). As shown in the formula for Hooke's Law for Ideal Elastic Solids (3.1), the proportionality constant is the modulus (G), which is a measure of stiffness and is expressed in units of dynes/cm^2.

$$S = G\gamma \tag{3.1}$$

The elongation resulting from the stretching of an elastic solid is compensated for by a decrease in the cross sectional area of the plastic specimen. This ratio of increase in length to decrease in area, called Poisson's ratio, is equal to 0.5 for ideal elastic solids. This value is approached for plasticized PVC or elastomers and decreased toward a value of 0.2 as less plasticizer is used in PVC or the crosslink density of the rubber is increased by reaction with more sulfur.

The bulk modulus (K) is the reciprocal of compressibility and is used as the proportionality constant in the expression for elastic solids in compression. In this case, the stress on an ideal solid with an initial volume of V_0 will be equal to the increase in pressure (ΔP) and the strain will be equal to the decrease in volume (ΔV) (3.2).

The relationship of bulk modulus to changes in pressure and volume:

$$K = \frac{\Delta P}{\Delta V / V_0} = \frac{\Delta P V_0}{\Delta V} \tag{3.2}$$

When the change caused by an applied force is the same in all directions, the solid is said to be isotropic. When this is not the case, as in some fiber-reinforced composites, the plastic is said to be anisotropic.

The irreversible flow of the viscous component of a plastic is related to the slippage of the polymer chains. According to the law for ideal liquids developed by Newton in 1685, the shear stress (S) is proportional to the velocity gradient or rate of shear ($d\gamma/dt$ or $\dot{\gamma}$). The proportionality constant as expressed in Newton's law for ideal liquids (3.3) is the coefficient of viscosity (η) expressed in units of poise or centipoise.

$$S = \eta \frac{d\gamma}{dt} \tag{3.3}$$

As shown by the Arrhenius equation (3.4) the viscosity (η) of a molten polymer decreases when the Kelvin temperature (T) is increased. In this equation, A is the proportionality constant, E is the activation energy required for flow, and R is the ideal gas constant.

$$\eta = Ae^{E/RT} \tag{3.4}$$

The resistance to flow (η) is related to the molecular weight (M). If one plots the logarithm of η against the logarithm of the average molecular weight ($\overline{M_w}$), one obtains a plot in which the log of viscosity increases linearly up to a break point at which the slope increases to about 3.4. The break point at which the slope increases represents the threshold molecular weight required for chain entanglement. The relationship of flow to molecular weight of polymers (3.5) is then expressed:

$$\eta = kM_w^{3.4} \tag{3.5}$$

All synthetic polymers and all natural polymers except proteins DNA and RNA consist of a mixture of molecules of different molecular weights and are said to be polydisperse. The number average molecular weight ($\overline{M_n}$) is an arithmetical average, whereas the weight average molecular weight ($\overline{M_w}$) is a larger value that is biased in favor of the higher molecular weight molecules. $\overline{M_w}$ is equal to $\overline{M_n}$ when the polymers are monodisperse (i.e., they all have the same molecular weight).

The viscoelasticity of a polymer may be represented by combinations of a spring model for the elastic component and a shock absorber or dash pot model for the viscous component. The modulus of the spring is equal to G and the viscosity of the fluid in the dash pot is equal to η. As shown by the models in Figure 3.1, Maxwell has proposed a model with the spring and dash pot in series and Voigt has proposed a model with these elements in parallel. A better representation of the effect of stress on a viscoelastic plastic is provided by using a combination of both models in series.

Plastics exhibit creep or irreversible changes in dimensions when subjected to stress over long periods of time. Likewise, they exhibit stress relaxation when stretched over long periods of time. Thus creep and stress relaxation are both time and temperature dependent.

It is not easy to measure viscoelastic properties such as creep and stress relaxation over the wide range of temperatures at which an amorphous

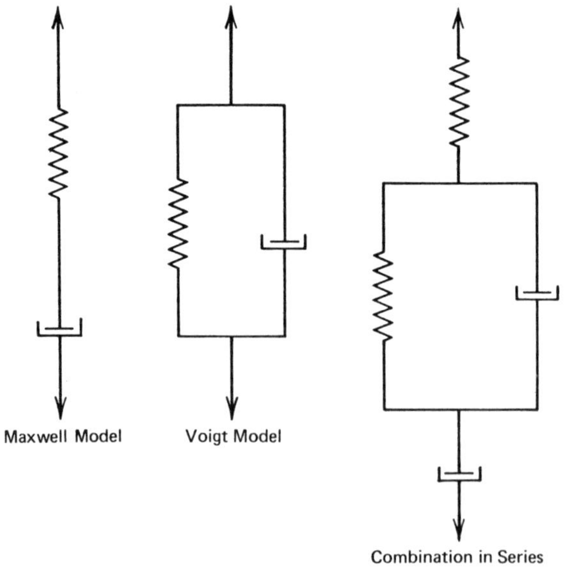

Figure 3.1. Maxwell and Voigt models.

plastic changes from a glassy solid to an elastomer to a liquid. However, Boltzman has developed time–temperature superposition relationships that provide useful master curves.

As shown by the Williams, Landel, and Ferry (WLF) equation (3.6),[8] one may use a shift factor a_T calculated relative to a reference temperature such as T_g to estimate the modulus of pure amorphous plastics or mixtures of plastics at another temperature T within 100 K of the reference temperature.

$$\log a_T = \frac{\log T(t)}{\log T_g(t)} = \frac{-17.44(T - T_g)}{51.6 + (T - T_g)} \tag{3.6}$$

Mechanical failure may occur under conditions of high stress for short periods of time, under conditions of low stress for long periods of time, and at any condition between these extremes. The stress–strain relationship (e.g., the tensile strength–elongation relationship of a plastic) may be described by the diagram shown in Figure 3.2.

In the generalized stress–strain curve, the stress, before the yield points, is Hookean and reversible. This elastic deformation is the result of instantaneous bending and stretching of the covalent bonds in the backbone of the polymers. There may also be a small amount of reversible uncoiling and disentanglement of polymer chains before the yield point.

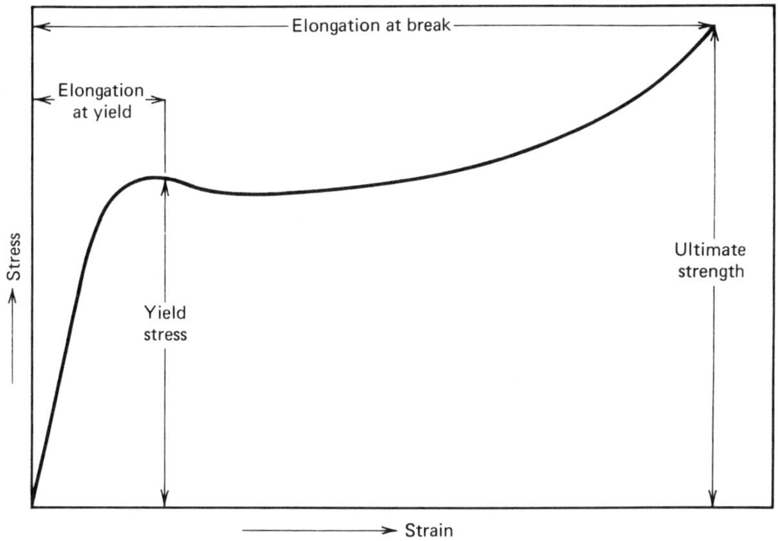

Figure 3.2. Generalized stress–strain curve for plastics.

There is complete disentanglement and irreversible slipping between polymer chains after the yield point, that is, in the plastic range. The extent of slippage is governed by the relative strength of the intermolecular forces present in the amorphous polymer. The area under stress–strain curves is a measure of toughness or resistance to impact.

As shown in Figure 3.3, Carswell and Nason[9] have used five curves to illustrate the stress–strain behavior of plastics. The soft and weak low-modulus–low-yield stress curve with moderate elongation at break is typical of a polymer such as polyisobutylene. The soft and tough low-modulus–low-yield stress curve, with high elongation at break is typical for a polymer such as plasticized PVC. The hard and strong high-modulus–high-yield stress curve, with moderate elongation at break is typical for a polymer such as rigid PVC. The hard and tough high-modulus–high-yield stress curve with high elongation at break is typical for a polymer such as ABS plastics. The hard and brittle high-modulus, very low-yield stress curve with low elongation at break is typical of a polymer such as polystyrene.

It is important to remember that the rate of the application of stress and temperature will change these descriptions by Carswell and Nason. The stress–strain curve of polystyrene at high temperature with slow application of stress will resemble that of polyisobutylene tested under normal conditions.

The modulus or rigidity of some polymers such as PVC can be reduced and the toughness increased by the addition of plasticizers. Thus Hyatt plasticized cellulose nitrate with camphor and Semon plasticized PVC with

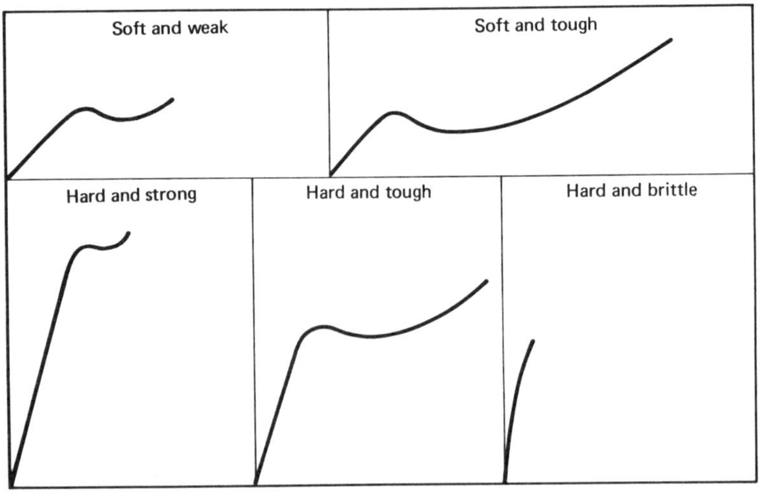

Figure 3.3. Typical tensile stress–strain curves for plastics.

tricresyl phosphate. The ductility of brittle plastics may also be improved by blending with more flexible polymers. For example, the brittle nature of PVC has been reduced by blending it with polyethyl acrylate.

In contrast, the modulus of plastics may be increased by the incorporation of fillers or reinforcing agents. Inasmuch as stress on these composites is transferred from the plastic (resin) or continuous phase to the reinforcing agent or discontinuous phase, it is essential that there be a good interfacial bond between the resin and the surface of the reinforcing agent. In some instances, this property is enhanced by the use of coupling agents such as organosilanes or organotitanates.

The physical properties of plastics are usually between those of fibers and elastomers. These properties, when measured under similar conditions of time and temperature, will vary in accordance with the intermolecular forces, the stiffening groups in the backbone, the pendant groups present on the chain, and the extent of crystallinity of the polymer.

Intermolecular forces are greatest in plastics such as polyamides, polyesteres, and polyurethanes, where hydrogen bonding provides strong intermolecular forces. The forces are moderate in plastics such as PVC, where the principal intermolecular forces are the result of dipole–dipole interaction. The weakest intermolecular bonds are the London or dispersion forces present in plastics such as polyethylene and polypropylene.

Phenylene, amide, and sulfone groups in the backbone act as stiffening groups. Ether, ester or methylene groups serve as flexibilizing groups. As shown in Table 3.1, these effects are additive so that the melting point T_m is decreased and the flexibility increased as the number of methylene groups in the repeat unit increases.

The pendant groups, when present on alternating carbon atoms and having a symmetrical arrangement in space, will increase the stiffness

Table 3.1. Effect of groups on the backbone on T_m

Plastic	Repeat Unit	T_m, °K
Polyethylene	$\{(CH_2)\}_n$	400
Polyether	$\{(CH_2)_3—O\}_n$	308
Polyamide	$\{(CH_2)_2—CONH\}_n$	598
Polysulfone	$\{(CH_2)_3—SO_2\}_n$	573
Aromatic polyester	$[(CH_2)_2—OCO—C_6H_4—COO]_n$	538

and the melting point of polymers. For example, isotactic or syndiotactic polypropylene, in which the methyl groups are all on the same side of the backbone or are arranged in an alternating order, is a rigid polymer that is not softened by exposure to steam. In contrast, atactic polypropylene, in which there is a random arrangement of the methyl pendant groups in space, is a viscous liquid at room temperature.

The flexibility will usually increase and the T_m decrease as the size of the pendant groups increases. Thus polybutyl methacrylate is more flexible than polymethyl methacrylate. However, larger well-ordered groups, such as those in isotactic and syndiotactic polymers, will cause an increase in melting point. Thus poly(3-methylpentene-1), which has an isobutyl pendant group, has a higher melting point than polypropylene.

The rigidity of polymers of dienes such as polyisoprene will depend on the arrangement of the bonds on each side of the ethylene carbon atoms. Those with a *cis* arrangement (i.e., bonds on the same side of the double bonds in the polymer), will be flexible, whereas those with a *trans* arrangement (i.e., bonds on opposite sides of the double bonds), will be rigid. The natural rubber elastomer is a poly *cis*-isoprene, whereas the hard higher melting gutta percha plastic is a poly *trans*-isoprene.

Although plastic with a low crosslink density (i.e., with few crosslinks) may be essentially as flexible as the corresponding linear polymer, those thermoset plastics with high crosslink density are usually intractable, insoluble, and not softened by heat. Ordinary vulcanized elastic rubber has a low crosslink density, whereas hard rubber or ebonite has a high crosslink density.

Additional information on the structure of polymers may be found in the standard polymer textbooks listed at the end of this chapter.[10–15]

REFERENCES

1. F. W. Billmeyer, *Textbook of Polymer Science,* 2nd ed., John Wiley & Sons, New York, 1971.
2. J. Brandrup and E. H. Immergut, *Polymer Handbook,* John Wiley & Sons, New York, 1975.
3. J. M. G. Cowie, *Polymers: Chemistry and Physics of Modern Materials,* Intext Educational Publishers, 1973.
4. J. P. Flory, *Principals of Polymer Chemistry,* Cornell University Press, Ithaca, NY, 1973.
5. H. S. Kauffman and J. J. Falcette, *Introduction to Polymer Science and Technology,* SPE Monograph, John Wiley & Sons, New York, 1977.
6. R. N. Lenz, *Organic Chemistry of Synthetic High Polymer,* Interscience Publishers, New York, 1967.
7. G. Odian, *Principles of Polymerization,* McGraw-Hill, New York, 1970.

8. F. Rodriquez, *Principles of Polymer Systems,* McGraw-Hill, New York, 1971.
9. R. B. Seymour, *Introduction to Polymer Chemistry,* McGraw-Hill, New York, 1971.
10. R. B. Seymour, *Introduction to Polymer Chemistry,* American Chemical Society, Audio Course, Washington, DC, 1980.
11. R. B. Seymour, *Modern Plastics Technology,* Reston Publishing Co., Reston, VA, 1975.
12. R. B. Seymour and C. E. Carraher, *Polymer Chemistry: An Introduction,* Marcel Dekker, New York, 1981.
13. M. P. Stevens, *Polymer Chemistry,* Addison-Wesley Publishing Company, Reading, MA, 1975.
14. B. Vollmert, *Polymer Chemistry,* Springer-Verlag Berlin, 1973.
15. H. L. Williams, *Polymer Engineering,* Elsevier Publishing Co., New York, 1975.

CHAPTER 4

Relationship of Molecular Structure to Chemical Resistance

As stated in Chapter 2, polymers consist of long chains of atoms in which the bond lengths, bond energies, bond angles, and other fundamental structural characteristics are similar to those in the repeat units in the polymer molecule. Although there may be a thousand or more of these repeat units joined together by covalent bonds in a polymer chain, the chemical properties of the accessible polymer molecules are similar to those of the repeat units or low molecular weight analogues (i.e., model compounds).

As a result of classical investigation of polyesterification reactions, Nobel laureate Flory has postulated an equal reactivity principle.[1] Flory showed that the rates of reaction of carboxyls with hydroxyl groups are independent of the size of the organic acid and alcohol, and all esterifications are third-order reactions. Thus as shown in (4.1), in the absence of inorganic acids the rate of esterification is proportional to the concentration of the alcohol and the square of the concentration of the organic acid.

$$\text{Rate of esterification} = k[\text{R'OH}][\text{RCOOH}]^2 \tag{4.1}$$

When oxygen, nitrogen, or sulfur are attached to the carbon atom, as in esters $(-\overset{\overset{\displaystyle O}{\|}}{C}-O-)$, acetals $(-O-\overset{\overset{\displaystyle H_2}{|}}{C}-O-)$, amides $(-\overset{\overset{\displaystyle O}{\|}}{C}-\overset{\overset{\displaystyle H}{|}}{N}-)$, or thioethers $(-\overset{\overset{\displaystyle H_2}{|}}{C}-S-\overset{\overset{\displaystyle H_2}{|}}{C}-)$, the bonds to the carbon atoms are more readily cleaved than when carbon atoms are present exclusively.

The continuation of a reaction of an amorphous polymer with a corrosive is dependent on diffusion of the attacking reactant. If the polymer is at a temperature above its T_g, there is a constant cooperative segmental movement that favors diffusion from hole to hole in the polymer mass.

The case of hole formation in the polymer above T_g is related to the relative mobility of the diffusing molecules and the chain segments. The rate of absorption of a corrosive molecule at temperatures below T_g is dependent to a large extent on the rate of change in the conformations of the polymer. In any case, the rate of diffusion is dependent on the temperature, the molecular size, and the extent of compatibility of the corrosive molecules with those of the polymer.

Although the equal reactivity principle can be used as a guide for the rate of reaction of accessible groups, the reactions with solid polymers may be hindered or enhanced by the presence of neighboring groups.[2] Reactions are hindered by crystallinity and enhanced by diffusion of the reactant into the polymer.

The electrical and thermal, as well as the chemical properties of polymers, are dependent on the chemical structure of the repeat unit. However, the mechanical properties of polymers are dependent on the size and morphology, or physical structure, of the polymer. The relationship of molecular structure to its chemical resistance for each type of polymer will be discussed in this chapter.

POLYALKENES

Polyalkenes such as high density polyethylene (HDPE) are the simplest polymers from a chemical viewpoint. The backbone of this linear polymer, as discussed in Chapter 2, corresponds to the backbone of many vinyl polymers. When heated in air, this polymer and most organic polymers will burn to produce carbon dioxide and water. However, this rapid type of oxidation is not typical in corrosive environments. Nevertheless, oxidized polyethylene surfaces which are antistatic, hydrophilic, and printable are produced commercially by corona discharge or treatment with hydrogen peroxide, nitric acid, or sodium hypochlorite.

The rate of oxidation of amorphous polyethylene is diffusion controlled and occurs faster than that of the crystalline polymer.[3] Because of the presence of more branching, the rate of oxidation of LDPE is faster than that of HDPE. The oxidation of atactic polypropylene is faster than that of isotactic PP and that of poly4-methyl-1-pentene (TPX) is much faster than that of PP. This increased rate is probably the result of increased diffusion rates because of the larger pendant groups present (4.2) as shown in these segments of typical polyolefins.

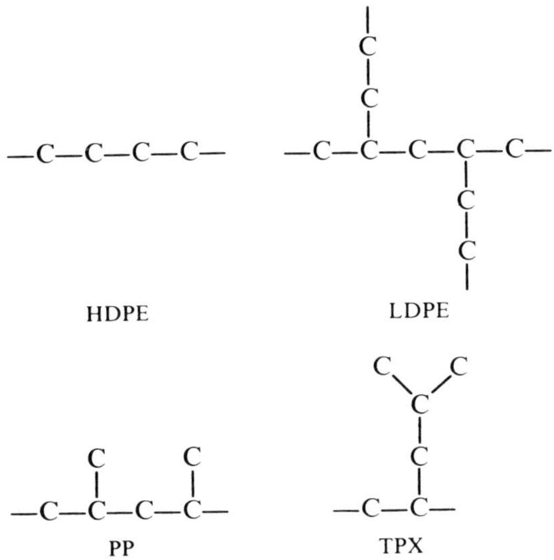

(4.2)

Oxidation of HDPE in the presence of sunlight is an autocatalytic free radical process similar to that observed for smaller alkanes or model compounds. As stated in Chapter 2, the bond energies of the carbon–carbon and carbon–hydrogen bonds are less than 100 kcal $mole^{-1}$. The wavelength of radiation in the near-ultraviolet region is as low as 290 nm, which corresponds to 100 kcal $mole^{-1}$.

It is important to note that much of the ultraviolet radiation energy absorbed by pure polyalkenes is dissipated without any bond breakage. However, when irradiated by UV radiation, polyalkenes degrade more readily when impurities are present. Of course, this effect is lessened in the presence of ultraviolet stabilizers. Because of the presence of a tertiary carbon atom, branched polyethylene (LDPE) and polyolefins with pendant groups degrade more rapidly than HDPE.

It is generally agreed that a free radical reaction is essential for this oxidation. The free radical may be produced by the cleavage of a carbon– carbon bond, by the decomposition of a hydroperoxide, or by direct oxidation of an alkane as shown in the following equations.

Cleavage of a carbon–carbon or carbon–hydrogen bond:

$$R\!:\!R \xrightarrow{E} 2\,R\cdot$$
$$R\!:\!H \xrightarrow{E} H\cdot + R\cdot \qquad (4.3)$$

Decomposition of a hydroperoxide:

$$ROOH \xrightarrow{E} RO\cdot + \cdot OH \tag{4.4}$$

Oxidation of an alkane (RH):

$$RH + O_2 \xrightarrow{E} R\cdot + \cdot OOH \tag{4.5}$$

As shown by the following equations, the alkyl-free radical R· may abstract a hydrogen atom from a polymer and produce a new free radical R′· or it may react with oxygen to produce a peroxy-free radical which in turn may abstract a hydrogen atom from a polymer.

The abstraction of a hydrogen atom from a polymer by an alkyl free radical:

$$\begin{aligned} RO\cdot + R'H &\longrightarrow ROH + R'\cdot \\ HO\cdot + R'H &\longrightarrow HOH + R'\cdot \\ R\cdot + R'H &\longrightarrow R'\cdot + RH \end{aligned} \tag{4.6}$$

The formation of a peroxy-free radical:

$$R\cdot + O_2 \longrightarrow ROO\cdot \tag{4.7}$$

The abstraction of a hydrogen atom from a polymer by a peroxy-free radical:

$$ROO\cdot + R'H \longrightarrow R'\cdot + ROOH \tag{4.8}$$

Although direct chain scission occurs in the presence of ultraviolet light or high temperature, the decomposition of the hydroperoxide groups as shown in (4.4) is responsible for much of the chain scission that occurs. Some crosslinking takes place in these oxidative processes. Hydroxyl, carbonyl (C=O), and carboxyl (COOH) groups have been identified in oxidized linear polyolefins, and ester groups have been also identified in oxidized crosslinked polyolefins.

Alkane polymers such as HDPE like their low molecular weight alkane analogues, react with chlorine at elevated temperatures or in the presence of ultraviolet radiation. Although chlorine may cause the degradation of polyalkene structures, pipe, or coatings, the chlorination process is used commercially to prepare chlorinated polyethylene (Tyrin).[5] As shown by the following equations, chlorination of alkanes, like the autoxidation reaction, is a free radical reaction in which the chlorine-free radical abstracts a hydrogen atom from the alkane.

Formation of chlorine-free radicals:

$$\mathrm{Cl:Cl} \xrightarrow{\mathrm{E}} 2\,\mathrm{Cl}\cdot \tag{4.9}$$

Abstraction of hydrogen atom from an alkane by chlorine-free radical:

$$\mathrm{Cl}\cdot + \mathrm{H:R} \longrightarrow \mathrm{H:Cl} + \mathrm{R}\cdot \tag{4.10}$$

Formation of chlorine-free radical by abstraction of chlorine by an alkyl-free radical:

$$\mathrm{R}\cdot + \mathrm{Cl:Cl} \longrightarrow \mathrm{R:Cl} + \mathrm{Cl}\cdot \tag{4.11}$$

Sulfonation of alkane polymers such as HDPE by chlorine and sulfur dioxide or sulfuryl chloride is also a free radical chain reaction. This reaction is also used to produce a commercial product called by the trade name Hypalon. Alkane polymers, like their low molecular weight analogues, will also react with nitric acid at elevated temperatures to produce nitroalkanes. Much bond cleavage occurs in this energy-rich reaction.

Since the strength of the bonds in fluorine molecules is much less than that in chlorine molecules, fluorine reacts readily with polyalkenes such as polyethylene at moderate temperatures in the absence of sunlight.[6] Dilute fluorine may penetrate deeply or simply produce a fluorinated polyalkene surface,[7] but inasmuch as fluorination is an energy-rich reaction, disintegration of the polymer usually occurs. Such disintegration is the result of cleavage of carbon–carbon bonds in the polymer backbone.[8]

It is important to point out that all of the reactions cited for polyalkenes will also take place with the methylene groups in other organic polymers. However, reactions with other reactants may take place more readily in a corrosive environment. In any case, the polyalkenes, namely, LDPE, HDPE, PP, polybutylene, polyisobutylene, and polymethylpentene, as well as copolymers such as those of ethylene and propylene or ethylene and 1-hexene, are resistant to most other corrosives not mentioned in the preceding discussion.

POLYARYLENES

Although polyarylenes such as polystyrene (PS) have been plagued by yellow discoloration and crazing in the past, these deficiencies have been overcome to a large extent by eliminating oxidized products and by reducing the concentration of residual monomer. Actually, pure PS remains colorless for long periods of time in the presence of ultraviolet light and autoxidation takes place less readily with PS than with LDPE.[9]

Thermal degradation of PS in an inert atmosphere occurs at about 280°C and the loss of volatile degradation products increases as the temperature is raised. The degradation and decrease in molecular weight of PS has been attributed to a scission of weak links in a reaction requiring 2000 propagation steps.[10]

These weak links have been attributed to oxygen links in the polymer. Both water and hydrogen peroxide are products of the degradation but not all weak links are cleaved.[11] The unzippering of poly-alpha-methylstyrene occurs at 61°C but the random thermal decomposition of PS does not occur below 280°C.

Chlorine, in the presence of sunlight or high temperature, will replace the hydrogen atoms on the backbone of PS. However, chlorination of the benzene ring does not occur readily in the absence of metallic impurities such as iron or iron chloride.[12]

Concentrated sulfuric acid will also attack the phenyl groups in PS. The original sulfonated polystyrene produced tends to crosslink by formation of sulfone groups. This sulfonation reaction, shown in (4.12), is facilitated by the presence of solvents.[13] Sulfonated PS is used commercially as an ion exchange resin.

Sulfonation of polystyrene:

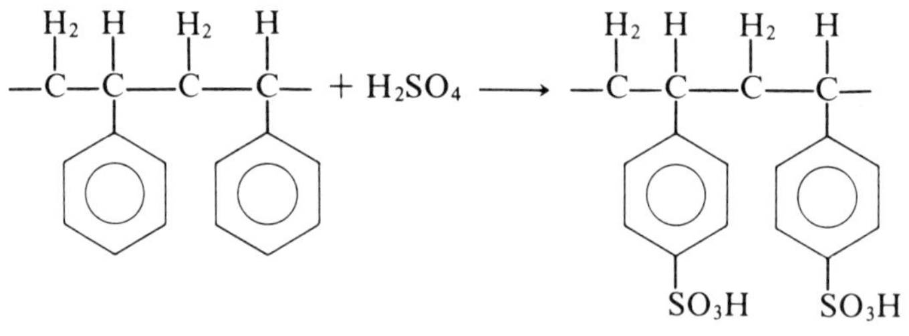

(4.12)

As shown in (4.13), polystyrene is readily nitrated by hot fuming nitric acid.[14] Mixtures of nitric acid and sulfuric acid will nitrate polystyrene and the product will dissolve in this acidic solution.[15] Nitrated polystyrene is readily reduced by heating with ammoniacal sodium hydrosulfite and the polyaminostyrene produced is used as an ion exchange resin.

Nitration of polystyrene:

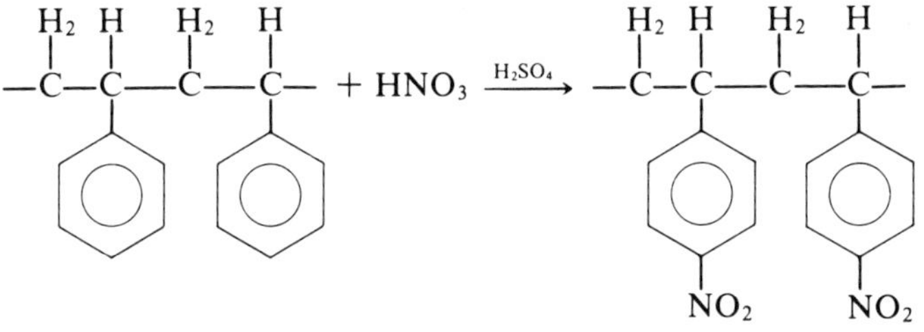

(4.13)

Sulfonation and nitration will also take place when copolymers of styrene such as the copolymer of styrene and acrylonitrile (SAN) (4.14), the block copolymer of styrene and butadiene (Kraton and Solprene), or ABS (acrylonitrile–butadiene–styrene) copolymers are exposed to hot sulfuric or nitric acids. These reactions will also take place with other polymers containing a benzene ring such as alkyds, diallyl phthalate, epoxy, phenolic, polycarbonate, aromatic polyester, polyphenylene oxide, polyphenylene sulfide, and polyphenyl sulfone.

Styrene-acrylonitrile copolymer:

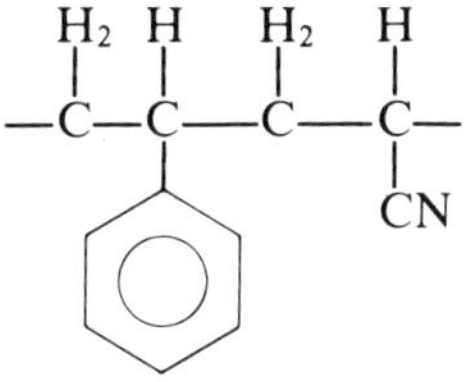

(4.14)

It is important to note that the reactions of copolymers are a function of each of the repeat units in the chain. Thus even acids such as hydrochloric acid, which does not affect PS, may hydrolyze the cyano group in ABS and produce carboxyl groups.

The rate of these reactions, including polymer degradation reactions, is accelerated by an increase in temperature. As shown by the Arrhenius equation (4.15), the rate constant k, which is related to the rate of the reaction, will be about double for every 10°C increase in temperature. As stated in Chapter 2, A is a frequency factor, E is the activation energy, R is the ideal gas constant, and T is the kelvin temperature. It is important to note that the viscosity decreases as the temperature increases and therefore the exponent has a positive sign. However, the rate constant k for chemical reactions increases and therefore the exponent for this Arrhenius equation is negative.

Arrhenius equation for rates of reaction:

$$k = Ae^{-E/RT} \tag{4.15}$$

In general, because of the absence of reactive functional groups, polyarylenes, like polyalkenes, are resistant to the attack of many corrosives at moderate temperatures.

DIENE POLYMERS

When heated at elevated temperatures, polymers of dienes such as natural rubber (polyisoprene), styrene–butadiene copolymer (SBR), and neoprene (polychloroprene) are depolymerized to produce the repeat unit diene and other decomposition products. The mechanism for autoxidation of diene polymers is similar to that described for polyalkenes.

Diene polymers are also degraded by ozone as the result of cleavage of the ozonides formed by the reaction of ozone and an ethylenic group. This degradative reaction (4.16) was used to determine the structure of natural rubber.[16]

Degradative ozonolysis of natural rubber:

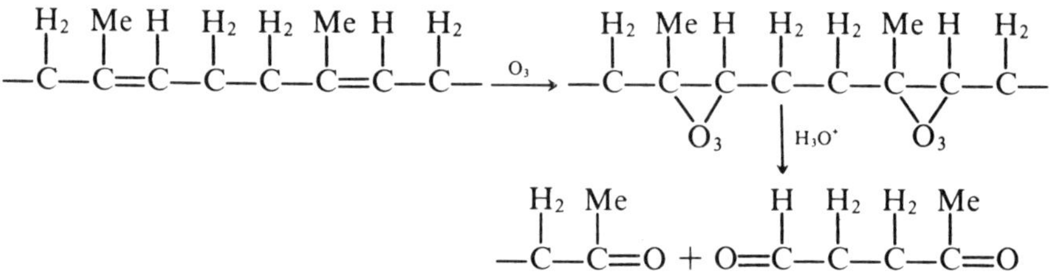

(4.16)

As demonstrated by the cracking of the side walls of tires in smog areas, exposure to ozone causes the cracking of rubber.

Chlorine also adds to the ethylenic groups in diene polymers and produces a nonelastic hard plastic.[17] Chlorinated rubber under the trade names of Tornesit and Parlon has been used as a base for protective coatings. The addition reaction is followed by cyclization and crosslinking of rubber and other diene polymers. Chlorine adds to the few double bonds present in butyl rubber and the product may be cured by the addition of metal oxides.

Hydrochloric acid also adds to diene polymers to produce rubber hydrochloride, a hard plastic.[18] This reaction is also accompanied by cyclization.[19] Rubber hydrochloride is used as a packaging film.

The reaction discussed for diene polymers takes place to a limited extent with synthetic polymers such as butyl rubber, and copolymers of ethylene and butadiene (EPDM) which have only a few ethylenic groups in each molecule. However, these reactions take place readily with unsaturated nonelastomeric diene polymers such as gutta percha and balata.

CHLORINE AND FLUORINE CONTAINING POLYMERS

Hydrogen chloride is produced when chlorine-containing aliphatic polymers such as PVC or polyvinylidene chloride (Saran) are heated at 150°C in the absence of air.[20] The rate of this allyl-activated dehydrochlorination is accelerated in the presence of oxygen.[21] PVC also becomes discolored when exposed to sunlight, changing from yellow to orange, brown, and black because of the chromophoric conjugated double bond produced by the dehydrochlorination reaction (4.17).

Thermal dehydrochlorination of PVC:

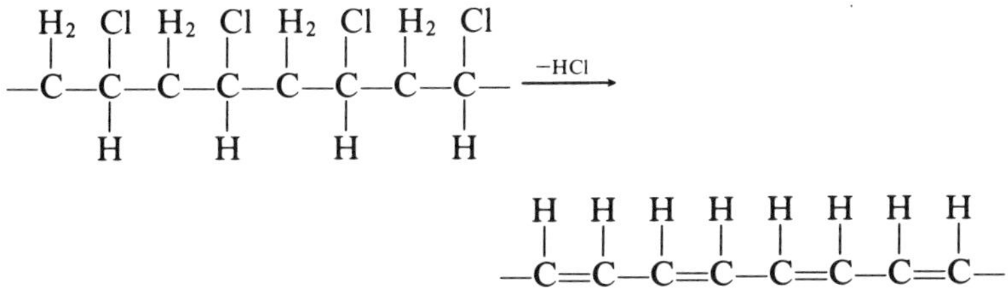

(4.17)

Chlorination of PVC produces a more highly chlorinated polymer (PVDC) which is more heat resistant than PVC. Polytetrafluoroethylene (PTFE) is one of the most stable polymers known. It is resistant to most solvents and corrosives at elevated temperatures (200°C). Fluorinated ethylene-propylene copolymer (FEP), a moldable copolymer of tetrafluoroethylene and hexafluoropropylene which also consists of only carbon and fluorine atoms, has similar resistance to solvents and corrosives. Polychlorotrifluorethylene (PCTFE), like FEP, is moldable and extrudable. It is slightly less resistant to corrosives and solvents than PTFE.

Polyvinylidene fluoride (PVDF), polyvinyl fluoride (PVF), and ethylene-tetrafluorethylene copolymer (ETFE) are less resistant to solvents and slightly less resistant to strong oxidizing agents than PTFE. However, these fluoroplastics do not lose hydrogen fluoride readily when heated and are not attacked readily by common corrosives.

HYDROXYL CONTAINING POLYMERS

Polymers with many hydroxyl groups such as polyvinyl alcohol (PVA), starch, and cellulose have limited use as corrosion materials. As demonstrated by the charring of cellulose, these polymers are readily dehydrated when heated. The hydroxyl groups are also oxidized when heated in the presence of oxygen and this reaction is accelerated in the presence of heavy metal salts such as iron or manganese compounds.

As shown by (4.18), the hydroxyl group is readily protonated and the protonated hydroxyl groups may lose water to form macrocarbonium ions. An anion from the acid may also add to the macrocarbonium ion to produce an unstable polymer, or the carbonium ion may lose a proton and form an unstable unsaturated polymer which is subject to further attack.

Protonation of a hydroxyl-containing polymer:

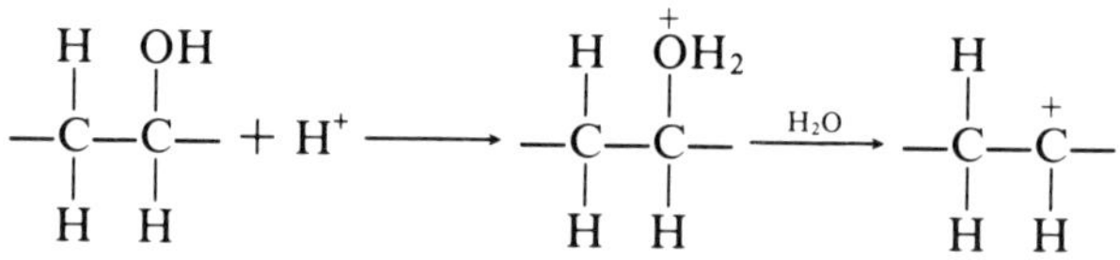

(4.18)

Addition of an anion from an acid to a macrocarbonium ion:

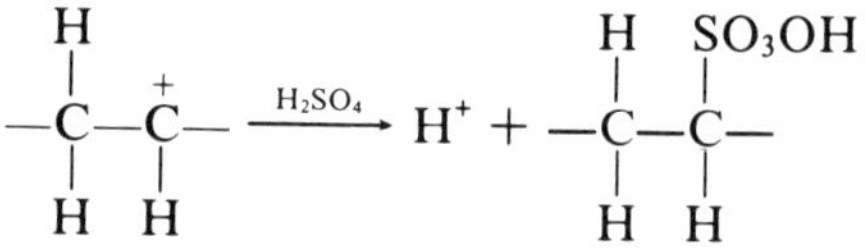

(4.19)

Deprotonation of a macrocarbonium ion:

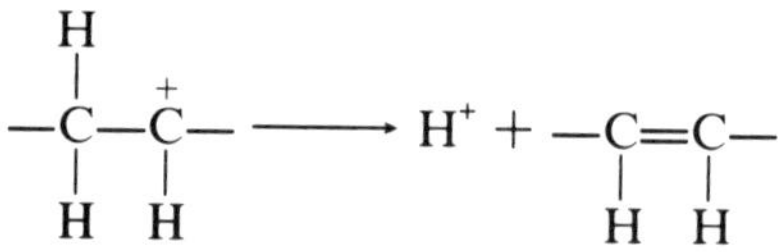

(4.20)

The reaction of sulfuric acid, sulfur dioxide, or sulfur trioxide with polymers containing hydroxyl groups yields esters. Reactions of cellulose with sulfuric acid have been investigated for many years.[22] Likewise, phosphates and nitrates are produced by the reaction of polymers containing hydroxyl groups with nitric or phosphoric acid.

It is believed that the nitronium ion (NO_2^+) is responsible for much of the nitration. This ion will even penetrate crystalline cellulose and produce cellulose nitrate.[23] The latter, of course, was used as one of the first man-made plastics. The extent of the reaction is a function of the concentration of the nitric acid. Some degradation also takes place when a hydroxyl containing polymer is heated with nitric acid or mixtures of this acid and sulfuric acid.

Acetic anhydride will acetylate hydroxyl groups in polymers and this reaction is catalyzed in the presence of inorganic acids such as sulfuric acid.[24] However, in the absence of mineral acids, organic acids have little effect on hydroxyl-containing polymers. Cellulose acetates with a degree of substitution of two and three are used commercially as plastics and fibers.

Hydroxyl-containing polymers form alkoxide ions when immersed in aqueous alkaline solutions. The formation of these anions, like the oxonium ions, formed in the presence of acids, reduces the hydrogen bonding forces and permits penetration of reactants into crystalline polymers. The immersion of cotton in aqueous sodium hydroxide solution was used to modify the water absorbency and dyeability of cotton by Mercer in 1850. The mercerization process is still in use today.

Inasmuch as sodium hydroxide solutions will dissolve cellulose, 17.5% sodium hydroxide has been used to separate lower-molecular-weight cellulose from the less soluble higher-molecular-weight fractions which are called alpha cellulose. The reaction of dimethyl sulfate or alkyl chlorides with alkali cellulose has been used to produce many cellulose ethers such as methyl, ethyl, and carboxymethylcellulose.[25]

Ethylcellulose is used as a melt coating, and methyl and carboxymethylcellulose (CMC) are used as water soluble polymers. Hydroxyl groups in hydroxyl-containing plastics also react with aldehydes such as formaldehyde and butyraldehyde. The acetal produced from the reaction of polyvinyl alcohol (PVA) and butyraldehyde is used as a clear plastic sheet in safety glass. The terminal hydroxyl groups in polymers of formaldehyde (acetals) and silicone polymers are converted to derivatives to assure thermal stability of these polymers.

The residual hydroxy groups in epoxy resins and alkyds are aliphatic hydroxy groups that may undergo the previously discussed esterification and esterification reactions. However, these hydroxyl groups in phenolic resins are acidic and form stable salts with alkalies.[26] Hence phenolic resin coatings or cements are not resistant to alkalies. Phenolic resins are resistant to 50% sulfuric acid but are attacked by oxidizing acids such as nitric acid.[27]

POLYESTERS

The hydrolysis of polyesters such as polymethyl methacrylate (PMMA), polyethyl acrylate (PEA), polyvinyl acetate (PVAc), alkyds, cellulose acetate

(CA), or esters of terephthalic acid such as polyethylene terephthalate (PET) or polybutylene terephthalate (PBT), in neutral or acid media involves a protonation of the ester. The initial reaction is followed by dealcoholation as shown by the following equations, in which the rate of hydrolysis is directly proportional to the concentration of water in the polymer.[28]

Protonation of esters in neutral or acidic medium:

$$R-\overset{\overset{O}{\|}}{C}-OR' \xrightleftharpoons{H^+} R-\overset{\overset{O}{\|}}{C}-\underset{+}{\overset{\overset{H}{|}}{O}}-OR' \qquad (4.21)$$

Dealcoholation of carbonium ions:

$$R-\overset{\overset{O}{\|}}{C}-\underset{+}{\overset{\overset{H}{|}}{C}}OR' \rightleftharpoons R-\overset{\overset{O}{\|}}{C^+} + R'OH \qquad (4.22)$$

$$R-\overset{\overset{O}{\|}}{C}-\underset{+}{\overset{\overset{H}{|}}{C}}OR' \xrightleftharpoons{H_3O^+} R-\overset{\overset{O}{\|}}{C}-O^+H_2 + R'OH \qquad (4.23)$$

These hydrolytic reactions cause cleavage of the polymer backbone in alkyds, PET, and PBT, but merely produce pendant carboxyl groups on PMMA and PEA, and pendant hydroxyl groups on PVAc and CA. The rate of hydrolysis is slowed by steric hindrance in aromatic polyesters. PMMA is much more resistant to hydrolysis than PEA.

Alkaline hydrolysis of solid polyesters (saponification) is slower than acidic hydrolysis because of the inability of the hydroxyl ions to penetrate the polymer.[29] Nevertheless, polyesters such as PET are more resistant to mineral or organic acids than to alkalies. It is important to note that the hydrolytic reactions cited for polyesters apply to copolymers such as block copolymers of polyesters and polyurethanes, and copolymers of vinyl acetate or methyl methacrylate with monomers such as vinyl chloride and styrene, respectively.

POLYAMIDES

Aliphatic polyamides such as nylon-66 and nylon-6 degrade when heated above 200°C or when exposed to ultraviolet radiation. This degradation is faster in the presence of oxygen[30] and metallic compounds including titanium dioxide.

These polyamides are hydrolyzed by hot water and dilute mineral acids such as 1*M* H_2SO_4 in a first order reaction in respect to the polyamide. As shown by the following, this hydrolysis involves a protonation of the amide followed by hydrolysis.[31]

Proposed mechanism for acid hydrolysis of polyamide:

$$\underset{\substack{\|\\ \mathrm{O}}}{\mathrm{R{-}C}}\mathrm{{-}}\underset{\substack{|\\ \mathrm{H}}}{\mathrm{N}}\mathrm{{-}R} \underset{}{\overset{H_3O^+}{\rightleftharpoons}} \mathrm{R{-}}\underset{\substack{\|\\ \mathrm{O}}}{\mathrm{C}}\mathrm{{-}}\overset{\substack{H_3O^+\\ |}}{\mathrm{N}}\mathrm{{-}R} \qquad (4.24)$$

$$\mathrm{R{-}}\underset{\substack{\|\\ \mathrm{O}}}{\mathrm{C}}\mathrm{{-}}\overset{\substack{H_3O^+\\ |}}{\mathrm{N}}\mathrm{{-}R} \overset{H_2O}{\rightleftharpoons} \mathrm{R{-}}\underset{\substack{\|\\ \mathrm{O}}}{\mathrm{C}}\mathrm{{-}OH} + H_2NR + H_3O^+ \qquad (4.25)$$

Acid hydrolysis is used to recover the reactants from scrap nylon-66.

Polyamides are also hydrolyzed in aqueous alkaline solutions but at a slower rate than in acid media. The aromatic polyamides (aramids) are more resistant to hydrolysis than the aliphatic polyamides.

POLYURETHANES

Polyurethanes are more resistant to hydrolysis than polyamides or polyesters.[32] The mechanism for attack by acids and alkalies is similar to that shown for polyamides.

POLYACETALS

Cellulose, starch, and their derivatives, as well as polymers of formaldehyde, are called polyacetals. Cellulose and starch are readily hydrolyzed to low-molecular-weight sugars by aqueous acid solutions or enzymes.[33] These hydrolytic reactions have been used as a source of glucose and ethanol.

Inasmuch as the esters are usually hydrolyzed, the hydrolytic products of cellulose esters are similar to those obtained from cellulose. However, hydrolysis of cellulose ethers yields ethers of glucose. Both cellulose and starch are more resistant to alkaline hydrolysis and higher temperatures must be used for hydrolysis at high pH's.[34]

The polyacetals (POM) are not attacked rapidly by corrosives in the pH range of four to nine at moderate temperatures. However, stronger acids or alkalies hydrolyze these polymers to formaldehyde. Because of the presence of phenylene stiffening groups, polyphenylene oxide (PPO), Noryl, and polyphenylene sulfide (PPS, Ryton) are more rigid than the compounding aliphatic polymers and are resistant to attack by corrosives.

The chemical resistance of plasticized plastics such as plasticized PVC, will be less than that of the unplasticized rigid polymer. In contrast, the chemical resistance of a reinforced plastic, such as fiberglass-reinforced polyester composites (FRP) will be greater than that of the unreinforced resin.

Chain symmetry, the presence of stiffening groups, and strong intermolecular secondary valence bonds will assure a high T_g and aid good packing of the polymer chains, thus contributing to chemical resistance. In contrast, the presence of randomly spaced pendant groups or flexibilizing groups which decrease T_g will also prevent good packing of the polymer chain and decrease chemical resistance.

Although there are many exceptions, one can usually predict resistance of polymers to corrosives from a knowledge of the structure of the repeat unit. Of course, when the selected polymer may be too expensive or too difficult to fabricate, the corrosion engineer must make compromises. Corrosion resistance tests and case history data will also aid the selection of a suitable plastic for a specific corrosive environment.

Additional data on organic chemistry may be found in references at the end of this chapter.[35–47]

REFERENCES

1. P. J. Flory, *J. Am. Chem. Soc.* **61,** 3334 (1939).
2. N. G. Gaylord, *J. Polymer Sci.* **24,** 1 (1968).
3. W. L. Hawkins, W. Matreyek, and F. H. Winslow, *J. Polymer Sci.* **41,** 1 (1959).
4. J. Bolland and G. Gee, *Trans. Faraday Soc.* **42,** 236 (1946).
5. E. W. Fawcett, U.S. Patent 2,183,556 (1939).
6. J. M. Tedder, in *Advances in Fluorine Chemistry*, M. Stacey, J. M. Tatlow, and A. G. Sharpe, Eds., Vol. II, Buttersworth, London, 1961.
7. J. Pinsky, A. Adylkonis and A. R. Nelson, *Mod. Packaging* **33**(6), 130 (1960).
8. J. J. Metz and R. B. Mesrobian, *J. Polymer Sci.* **16,** 45 (1955).
9. D. A. Anderson and E. S. Freeman, *J. Polymer Sci.* **54,** 253 (1961).
10. N. Grassie and W. W. Kerr, *Trans. Faraday Soc.* **55,** 1050 (1959).
11. H. H. G. Jellinek, *J. Polymer Sci.* **3,** 850 (1948).
12. G. B. Bachman, et al., *J. Org. Chem.* **12,** 108 (1947).

13. M. Baer, U.S. Patent 2,523,210 (1950).
14. J. Blyth and A. W. Hofman, *Ann. Chem.* **53,** 316 (1843).
15. H. Zeuftman, *J. Chem. Soc.* 982 (1950).
16. C. D. Harries, *Ber.* **37,** 2708 (1904).
17. G. A. Englehard and H. H. Day, British Patent 2734 (1859).
18. H. Staudinger, *J. Propt. Chem.* **162,** 148 (1943).
19. W. C. Calvert, U.S. Patent 1,989,632 (1935).
20. G. Ayrey, B. C. Head, and R. C. Poller, *J. Polymer Sci.* **D 8,** 1 (1974).
21. J. H. L. Henson and F. J. Hybart, *J. Appl. Polymer Sci.* **17,** 129 (1973).
22. H. Braconnot, *Ann. Chem. et Phys.* **12**(2), 185 (1819).
23. F. D. Miles, *Cellulose Nitrate*, Oliver and Boyd, London, 1955.
24. C. J. Malm, *Ind. Eng. Chem.* **28,** 1310 (1936).
25. J. K. Chowdhury, *Biochem. Z.* **148,** 76 (1924).
26. W. G. Simons, U.S. Patent 2,593,926 (1952).
27. R. B. Seymour, *Corrosion* **7**(5), 151 (1951).
28. D. A. S. Ravens, *Polymer* **1,** 385 (1960).
29. E. Waters, *J. Soc. Syers Colorists* **66,** 609 (1950).
30. G. I. Goodman, *J. Polymer Sci.* **13,** 175 (1954).
31. V. A. Myagkov and A. B. Pakshves, *Kolloidn Zk* **14,** 172 (1952).
32. E. R. Wells, V. C. Hixenbaugh, and G. A. Hudson, *Paint Varnish Prod.* **53**(8), 33 (1963).
33. J. H. S. Saunders, S. Steingeiser, et al., *J. Chem. Eng. Data* **3,** 153 (1958).
34. O. A. Battista, *Ind. Eng. Chem.* **42,** 502 (1950).
35. R. F. Brown, *Organic Chemistry*, Wordsworth Publishing Co., Belmont, CA, 1975.
36. W. H. Brown, *Introduction to Organic Chemistry*, Willard Grant Press, Boston, 1978.
37. J. M. Crain and D. M. Crain, *The Essence of Organic Chemistry*, Addison-Wesley, Reading, MA, 1918.
38. T. A. Geissman, *Principals of Organic Chemistry*, W. H. Freeman & Co., San Francisco, 1927.
39. H. Hart and R. D. Schuetz, *Organic Chemistry—a Short Course*, Houghton Mifflin, Boston, 1978.
40. J. R. Holum, *Organic and Biological Chemistry*, John Wiley & Sons, New York, 1978.
41. R. T. Morrison and R. N. Boyd, *Organic Chemistry*, Allyn & Bacon Inc., Boston, 1973.
42. R. B. Seymour, *General Organic Chemistry*, Barnes & Noble, New York, 1971.
43. R. B. Seymour and J. G. Higgins, *Experimental Organic Chemistry*, Barnes & Noble, New York, 1971.
44. T. W. G. Solomons, *Organic Chemistry*, John Wiley & Sons, New York, 1980.
45. M. P. Doyle and W. S. Mungall, *Experimental Organic Chemistry*, John Wiley & Sons, New York, 1980.
46. A. L. Ternay, *Contemporary Organic Chemistry*, W. B. Saunders Co., Philadelphia, 1979.
47. P. Wiseman, *An Introduction to Industrial Organic Chemistry*, Update Publishing International, Englewood, NJ, 1979.

CHAPTER 5

Effect of Solvents on Polymers

The theories on the degradation of polymers by corrosive electrolytes such as acids, bases, and aqueous salt solutions discussed in Chapter 4 are not as straightforward as theories for metallic corrosion. However, most of the theories for the effect of solvents on polymers are at least as useful as the theories on metallic corrosion. Much of our present knowledge of solutions is based on principles developed by Hildebrand and Scott in 1924.[1]

Fortunately, most solvents are nonpolar and therefore their intermolecular forces are weak London or dispersion forces. Hildebrand used the term "regular solutions" to describe solutions of nonelectrolytes and their nonpolar solvents. Additional theories on the solubility of polymers were developed by Flory[2] and Huggins.[3] Probably the most important publications leading to the practical use of solubility theories by polymer engineers were those published by Burrell starting in 1955.[4]

According to Hildebrand, the compatibility of a solvent and a nonelectrolyte solute is dependent on the magnitude of the change in enthalpy (ΔH) when these substances are mixed. The value of ΔH is dependent on the molar volume of the mixture (V), the mole fraction of the components (ϕ) and the difference in their cohesive energy densities (CED).

Because values for *CED*, the energy of vaporization per cm^3, were large and cumbersome, Hildebrand and Scatchard[5] used the square root of *CED*, called the solubility parameter (δ). These relationships are shown in (5.1) and (5.2).

Relationship of cohesive energy density to the solubility parameter:

$$CED = \frac{\Delta E}{V} = \delta^2 \quad \text{or} \quad \delta = \left(\frac{\Delta E}{V}\right)^{1/2} \tag{5.1}$$

Relationship of energy of mixing to the solubility parameter for regular solutions:

$$\Delta H = V\phi_1\phi_2(\delta_1 - \delta_2)^2 \tag{5.2}$$

When a solute dissolves, the value for energy of mixing (ΔH) is small and the value of $\delta_1 - \delta_2$ approaches zero. Conversely, if ΔH is very large, the solute will not dissolve. Thus an amorphous polymer will dissolve if the difference in the solubility parameters ($\delta_1 - \delta_2$) is small and the intermolecular forces (i.e., London or dispersion forces), are small.

As shown by the Gibbs' free energy equation (5.3), the change in free energy (ΔG) will be negative and equal to the product of the kelvin temperature (T) times the change in entropy (ΔS) when the change in enthalpy (ΔH) is equal to zero. The requirement for a spontaneous reaction (i.e., that ΔG be negative) will be met even when the value of ΔH is small, providing the value for $T\Delta S$ is greater than ΔH. Thus an increase in temperature T should increase the value of $-\Delta G$ and therefore increase the solubility of an amorphous polymer in a nonpolar solvent.

Gibbs' Free Energy Equation:

$$\Delta G = \Delta H - T\Delta S \tag{5.3}$$

Modifications in the Hildebrand solubility parameter concept for regular solutions to account for larger intermolecular forces were made by Lieberman,[6] Crowley,[7] and Hansen.[8] Although the resulting three-dimensional parameters are not in agreement, they do emphasize that modifications of the Hildebrand solubility parameter concept must be made for polar solvents or polar solutes.

In any case, the solubility parameter of a solvent may be estimated from the definition of *CED*. Thus, the solubility parameter δ is equal to the square root of $\Delta E/V$ as stated in (5.1). However, since $\Delta H = \Delta E + RT$, $\Delta E = \Delta H - RT$. Thus $\delta = (\Delta H - RT/V)^{1/2}$. As it is much easier to measure molecular weight M and density D, these values are substituted in the relationship of $V = M/D$.

Solubility parameter equation:

$$\delta = \left(\frac{\Delta H - RT}{M/D}\right)^{1/2} \tag{5.4}$$

The usefulness of this expression is illustrated by (5.5) in which the δ value for heptane at 298 K is calculated using values of $\Delta H = 8700$ cal/mol, $R = 2$ cal/mol-K, and $D = 0.68$ g/cm^3. The solubility parameter values for many solvents and polymers are available in tabular form and are shown in Tables 5.1 and 5.2.[9,10]

$$\delta = \left(\frac{\Delta H - RT}{M/D}\right)^{1/2} = \left(\frac{8700 - 2(298)}{100/0.68}\right)^{1/2} = 7.4H \tag{5.5}$$

Table 5.1. Solubility parameters of selected solvents[a]

Solvent	$\delta(H)$
Dimethylsiloxane	5.5
Difluorodichloromethane (Freon 12)	5.5
Diisopropyl ether	6.9
Hexane	7.3
Octane	7.6
Butyl chloride	8.1
Methyl amyl ketone	8.5
Toluene	8.9
Chlorobenzene	9.5
Acetone	9.9
Isobutyl alcohol	10.5
Nitroethane	11.1
Isopropyl alcohol	11.5
Dimethyl sulfoxide (DMS)	12.0
Ethanol	12.7
Methanol	14.5
Glycerol	16.5
Formamide	19.2
Water	23.2

[a]The strength of intermolecular bonding forces increase as δ values increase and must be considered.

Table 5.2. Approximate solubility parameter values of selected polymers

Polymer	$\delta(H)$
PTFE	6.2
Polydimethylsiloxanes (silicone)	7.3
Bitumens (asphalt)	7.5
Natural Rubber	8.0
HDPE	8.0
PP	8.4
SBR	8.5
PS	9.2
PMMA	9.2
PVC	9.4
Alkyds	9.4
PU	10.0
PET	10.5
Epoxy resins	10.9
Cellulose acetate	11.4
Nylon-66	12.0
PVA	12.6
Cellulose	15.6

The dimension of δ may be expressed as $(\mathrm{cal/cm^3})^{1/2}$. However, it is more convenient to express these parameters in Hildebrand units.[11]

Since polymers are not volatile, their solubility parameter values cannot be estimated from (5.4). These values can be estimated experimentally by using a spectrum of solvents and assigning the δ value of the solvent that yields a solution with the highest viscosity or by titrating with a poorer solvent.[12]

However, δ values for polymers may be estimated by use of molar attraction constants.[13] As shown in (5.6), Small obtained δ values by multiplying the sum of the molar attraction constants G by D/M.

Small's formula for solubility parameter:

$$\delta = \frac{D\Sigma G}{M} \tag{5.6}$$

Inasmuch as the molar attraction constants for methyl (CH_3), methylene (CH_2), and methine (CH) groups are 214, 133, and 28, respectively, one may calculate the solubility parameter of polypropylene (PP) which has a density of 0.905 g/cm^3 by using the molecular weight (42) and G values of the repeat unit [$-CH(CH_3)CH_2-$].

$$\delta = \frac{D\Sigma G}{M} = \frac{0.905(214 + 133 + 28)}{42} = 8.1H \tag{5.7}$$

Of course, inasmuch as isotactic PP is crystalline, it will not dissolve in cyclohexane ($\delta = 8.2H$) below T_m. However, atactic PP is soluble in cyclohexane. A linear and branched amorphous polymer will usually dissolve in solvents which have δ values within $1.8H$ of that of the polymer, but polymers will swell in solvents having δ values within $3.2H$ of that of the polymer.[14] Nevertheless, crosslinked or thermoset plastics will swell but will not dissolve in good solvents.

The solubility parameter provides a quantitative approach for the "like dissolves like" rule. Since the Hildebrand concept obeys the law of mixtures, δ values for mixture of miscible solvents may be calculated if the relative mole fraction of the solvents is known.

It is important to note that immiscible solvents act independently. Thus an aqueous emulsion of benzene ($\delta = 9H$) will dissolve polystyrene (PS) ($\delta = 9.2H$).[15] However, as demonstrated by the classical experiments of Menard, a mixture of nonsolvents will dissolve a polymer if the δ value of the mixture is similar to that of the polymer. Cellulose nitrate is insoluble in ethyl ether ($\delta = 7.4H$) and in ethanol ($\delta = 12.7H$) but is soluble in an equimolar mixture ($\delta = 10H$).

Plastics which neither dissolve nor swell in a solvent may crack in contact with the solvent. This "stress cracking" which was first described in 1951[16] has been the subject of many investigations.[17] This phenomenon was first noted when polar liquids were stored in LDPE bottles and has been labeled "environmental stress cracking" (ESC). The extent of cracking increases as the temperature is increased.

The phenomenon (ESC), which is seldom noted in the absence of stress, is a physical process which may occur occasionally with other polymers. ESC may be evaluated by use of surface active agents and is believed to be caused by the presence of low-molecular-weight polymers in the plastic. HDPE, ultra-high-molecular-weight LDPE, ethylene copolymers, and fiber-filled and annealed plastics have superior resistance to ESC.

Cracking of plastics may also occur with nonpolar solvents. Thus LDPE may crack in the presence of benzene or ethylene dichloride,[18] and PS may crack in the presence of aliphatic hydrocarbons or butter fat.[19]

As shown by the data in Tables 5.1 and 5.2, solvents and polymers have a wide range of solubility parameters. Thus if one knows these values, a polymer that will not be adversely affected by the solvent may usually be selected.

It is important that plastic materials of construction be solvent resistant. However, it is equally important that polymers used in solution be soluble. The guidelines of $\pm$ 1.8H is helpful in the selection of soluble plastics. Soluble polymers are required for many applications such as detergents, viscosity improvers, solution coatings, adhesives, enhanced oil recovery, flow enhancement, and thickeners.

The guidelines of $\pm$ 1.8H for solubility and $\pm$ 3.2H for swelling may also be used to select solvent-resistant polymers. Data from objective tests such as those discussed in Chapters 6 and 7 are essential for the final selection of plastics and other polymers for use in corrosive environments.

REFERENCES

1. J. H. Hildebrand and R. L. Scott, *The Solubility of Nonelectrolytes*, 3rd ed., Reinhold Corp., New York, 1950.
2. P. J. Flory, *J. Chem. Phys.* **9**, 660 (1941).
3. M. L. Huggins, *J. Chem. Phys.* **9**, 440 (1941).
4. H. Burrell, *Official Digest, Fed. Soc. Paint Technol.* **27**(369), 726 (1955), **38**(492), 43 (1966).
5. G. Scatchard, *Chem. Rev.* **8**, 321 (1931).
6. E. P. Lieberman, *Official Digest, Fed. Soc. Paint Technol.* **34**, 444 (1962).
7. J. D. Crowley, G. S. Teague, and J. W. Lowe, *J. Paint Technol.* **39**(492), 269 (1966).
8. C. W. Hansen, *J. Paint Technol.* **39**(501), 104 (1967).

9. H. Burrell and B. Immergut, Chapter 4–5 in *Polymer Handbook,* J. Brandrup and E. H. Immergut, Eds., Wiley-Interscience, New York, 1966.

10. R. B. Seymour, *Handbook of Chemistry and Physics,* 56th ed., C-732, 1979.

11. T. W. Harris and R. B. Seymour, *Structure-Solubility Relationship in Polymers,* Academic Press, New York, 1977.

12. R. B. Seymour and H. A. Wood, Chapter 8 in *Structure-Solubility Relationships in Polymers,* F. W. Harris and R. B. Seymour, Eds., Academic Press, New York, 1977.

13. P. A. Small, *J. Appl. Chem.* (3), 71 (1953).

14. R. B. Seymour, H. S. Tsang, E. E. Jones, A. K. Patal, and P. D. Kincaid, *Advances in Chemistry Series* **99,** 418 (1971).

15. R. F. Eisenberg, R. R. Kraybill, and R. B. Seymour, Chapter 1 in *Laboratory Engineering and Manipulations,* E. S. Perry and A. Weissberger, Eds., Wiley-Interscience, New York, 1979.

16. J. B. Decoste, F. S. Malm, and V. T. Wallder, *Ind. Eng. Chem.* **43,** 117 (1951).

17. J. B. Howard, Chapter 11 in *Engineering Design for Plastics,* E. Baer, Ed., Reinhold Publishing Co., New York, 1964.

18. F. J. Bockhoff and J. A. Neumann, *SPE J.* **10**(5), 17 (1954).

19. E. E. Ziegler and W. E. Brown, *Plastics Technol.* **1**(341), 409 (1955).

CHAPTER 6

Tests for Corrosive Resistant Coatings

As stated in Chapter 4, the degradation of plastics and other polymers in corrosive environments involves many different chemical reactions that eventually cause scission of the covalent bonds in the polymer backbone and degradation of the plastic. In contrast, the corrosion of metals is a natural chemical process in which the metal is oxidized through loss of electrons to a more stable state at a characteristic voltage.

CORROSION OF METALS

As shown by the following equation for the oxidation half-cell reaction, a metal such as iron is oxidized to a higher valence state when it is subject to corrosion. In the case of iron, it is oxidized to iron(II) ion or ferrous ion (Fe^{2+}) with a loss of two electrons.

Half-cell reaction for the corrosion of iron:

$$Fe_{(S)} \longrightarrow Fe^{2+} + 2e \qquad E = -0.4\ \text{V} \tag{6.1}$$

Metals above hydrogen in the electromotive series are readily oxidized by nonoxidizing acids. The tendency for such corrosion to take place is influenced by the hydrogen ion concentration, that is, pH, and the temperature.

Metals below hydrogen in the electromotive series are not attacked by nonoxidizing acids but they may be oxidized by nitric acid, concentrated sulfuric acid, and other oxidizing environments. The electrolytic corrosion process may be shown mathematically by the Nernst equation (6.2), where E_0 equals the standard electrode potential, R equals the ideal gas constant, T equals the kelvin temperature, n equals the number of electrons transferred per atom in the oxidation, F equals the Faraday, and K equals the equilibrium constant, [i.e., [product]/[reactants] for the oxidation–reduction reaction].

Nernst equation:

$$E = E_0 - \frac{RT}{nF} \ln K \tag{6.2}$$

The voltage for this oxidation reaction changes as the concentration of hydrogen ions decreases. The maximum value for the electrode potential in neutral water is −0.41 V. Even in a 1 *M* solution of iron, there is a driving force of 0.03 V favoring corrosion. The driving force will be greater in common corrosive environments. The tendency for corrosion to occur may also be shown by the change in Gibbs' free energy, ΔG (6.3).

Gibbs' equation for corrosion:

$$\Delta G = -RT \ln K \tag{6.3}$$

Inasmuch as the degradation of metals in a corrosive environment is an electrochemical process dependent on the concentration of hydrogen ions, the tendency for corrosion to take place may be predicted. Likewise, inasmuch as metal ions are removed at a constant rate, the loss of thickness of the test sample may be monitored by loss of thickness measurements. Thus the corrosion engineer can measure the rate of corrosion of metals in mm per year (mpy) and estimate the useful lifetime of a metal structure in a corrosive environment.

This degradative process may be prevented by using more active metals as sacrificial anodes, by imposing a higher voltage to prevent the electrochemical reaction through cathodic protection, and/or by protecting the metal surface by the application of a coating or lining. Although tests for the corrosion of metals are well defined, this is not true for the coatings or linings.

TESTS FOR COATINGS

Salt Spray Test

Prior to the development of the chemical process industry, considerable deterioration of metals occurred in soils, humid atmospheres, and coastal areas. Presumably, the latter appeared to be a severe test of any material of construction, and there appeared to be some chance of simulating the conditions of exposure.

Certainly one set of values under controlled conditions would be preferable to those noted for atmospheric exposure under varying weather conditions. Also dynamic tests apparently could be considered more significant than simple static immersion tests. In addition, the salt spray test, which is an accelerated test, served to meet the requirements of a standard exposure test.[1]

This accelerated test (ASTM B-117) has been widely employed as a basis for arbitrary performance standards in any specifications for coated metals. In this test, a specimen is suspended in a mist of an atomized aqueous solution of sodium chloride in a chamber at a specified temperature. The recommended exposure periods are 16, 24, 48, 96, 200, 240, 500, or 720 hours. The supplier and user may agree on other exposure times, but intervals of 24 hours are suggested by ASTM.

The specified conditions for the salt spray (fog) test usually are a pH of 6.5–7.2 at 35°C. However, a pH of 3.1–3.3 is used in the acetic acid–salt spray (fog) testing method (ASTM B-287). One automobile manufacturer's laboratory used the Cas test (ASTM-268) which utilizes a 5% sodium chloride solution containing 4 g/liter of copper(II) chloride at a pH of 3.2 at 50°C.[2]

The use of the salt spray test for questionable purposes has resulted in considerable controversy and misunderstanding. However, in spite of its limitations and unreliability, this test is still popular. Comparative data showing weight loss versus exposure time for test specimens using natural and synthetic seawater under different conditions have been published.[3] Presumably this test will continue to be considered even when better tests are developed.

OUTDOOR WEATHERING TESTS

Outdoor exposure tests for coated panels in Atlantic City were described by the American Association for Testing Materials over 70 years ago.[4] Since then, hundreds of thousands of coated panels have been exposed on racks in various locations, including the National Bureau of Standards,[5] at an angle of 45° to the horizontal. It is customary to place the panels facing south. Since these tests require long periods of time, they are often supplemented and sometimes replaced by accelerated tests.

ACCELERATED TESTS

In addition to the salt fog test mentioned previously, there are accelerated tests for the exposure of coatings to artificial light and water (ASTM D-822), to mirror-intensified sunlight (EMM AQUA),[6] and accelerated weathering tests (ASTM D-795).

The most comprehensive accelerated weathering test (ASTM-G-23) utilizes a weatherometer which includes a sunshine arc and humidity control. In this cyclic test, the test samples are exposed to simulated sunlight at 50% humidity for one hour and then to 100% humidity in the dark for another hour. Water cooled Xenon arcs and fluorescent light sources are used in the Xenotest and Climatest.[7] Cyclic heating and cooling as well as salt water wetting and drying have been proposed in an accelerated test for coatings in coastal areas.[8]

Static Immersion Tests

Immersion of coated test specimens under static conditions disregards many significant factors encountered under conditions of use. Fortunately, the volume of solution is not significant if it is large relative to the size of the test piece. Accordingly, it has been recommended that 20 to 200 ml of solution should be used for a test specimen with a surface area of 6.45 cm^2. The proportion of exposed edge section to the entire surface area should be small.

Tables in which the results of static immersion tests are recorded are widely used.[9] These ratings are often reported qualitatively as *A* for good, *F* for fair, *C* for conditional and *X* for unsuitable.

Immersion tests may be accelerated somewhat when care is used and limitations of the test are recognized. Standards have been developed for organic coatings on steel (ASTM D-870). In these tests, the partially immersed specimen is maintained at a temperature of 38 ± 1°C for the agreed upon length of time. Adjacent panels are separated by at least 1 cm and the test liquid is changed every 72 hours to minimize contamination.

Modifications of these tests include aeration[10] and alternate immersion (ASTM B-192, 870). In the latter type test, the sample may remain wet when out of the test solution or it may be dried by a blast of warm dry air.

The immersion tests may be used with water (ASTM D-870) or aqueous solution of acids, alkalies, and salts, as well as organic liquids. The ASTM Test lists 51 standard test reagents (ASTM D-543).

Other Test Environments

Considerable information on atmospheric testing of coated materials has been developed by paint technicians. For example, coated steel panels have been tested for outdoor exposure using test fences facing north and south with test specimens in a vertical position (ASTM D-1014).

Standard forms for recording results of exposure tests of paints when accompanied by photographs make such tests much more objective (ASTM D-1150). These forms permit the rating of change in properties with time using a rating of 1 to 10. Photographs of exposed panels may be compared with typical photographs included with the ASTM specifications.

This type evaluation has been referred to by use of ASTM photographic reference standards for the determination of the degree of resistance of paints to chalking, checking, cracking, erosion, and flaking (scaling). Photographic reference standards have also been used to evaluate resistance to rusting of iron surfaces protected by organic protective coatings (ASTM D-610). Although the emphasis has been on paints, the tests, of course, have been based on the rate of disintegration of thin plastic films. The procedures are empirical, yet can yield a certain degree of reproducibility.[11]

Some tests developed by the paint industry for use in quality control may be extended to the evaluation of environmental effects on plastic films. For example, the casting of films with a doctor's blade provides a basis for more objective tests. The quality control tests include flexibility tests in which a control panel is bent over a conical mandrel and the relative position of cracks is determined (ASTM D-552). The distance along the axis of the cone from its base is corrected for thickness by use of a graph and the percentage of elongation is estimated using the corrected value on another graph. Such tests may be made before and after exposure to corrosives. Competent technicians have revised empirical tests so that they have a reasonable degree of reproducibility and may be considered objective to some degree.

The capacitance and resistance of polybutadiene coatings on steel has been determined as a function of the time of exposure of the control specimens to air saturated with 3% sodium chloride. In this test, there was an abrupt change in resistance at the time of failure.[12]

The effectiveness of coatings has been estimated by measuring the change in voltage across electrodes on the surface of a coated panel,[13] by measuring the change in electrical resistance of a coated panel,[14] and by using the coated specimens as a cathode in a salt solution and measuring the current flow, if any.[15]

A laser interferometer technique has been used for the detection of holidays in coated metal samples.[16] As much as three times the base electrode potential was required for the same current flow with a coated steel specimen. Short-time electrical impedance tests in combination with classical hardness and adhesion tests have been proposed for testing barrier coatings.[17]

The change in thickness of a coating during exposure may be measured by determining the changes of the back scatter or radiation,[18] or by magnetic thickness tests. Changes in adhesion of a coating have been measured by using ultrasonics.[19]

Other procedures include measurements of changes in physical properties after exposure to corrosives (ASTM D-1308, D-1654), changes in hardness,[20,21] and changes in abrasion resistance.[22] Permeability of coatings has been evaluated by use of radiotracer techniques[23,24] and by measuring the rate of penetration through the film.[25]

FILIFORM CORROSION

In addition to attack by corrosion on organic coatings, we must also consider another destructive mechanism called filiform corrosion. This term was coined to describe an underfilm corrosion phenomenon first noted some 35 years ago.[26] This particular type of underfilm electrochemical corrosion takes place along specific threadlike patterns under high humidity conditions (ASTM D-2805) and is accelerated by the presence of sulfur dioxide and inhibited by carbon dioxide.[27]

It has been postulated that filiform corrosion results from the flow of corrosive currents through a film of electrolyte beneath the coating and that this flow is enhanced by the presence of contaminants that provide conducting ions.[28] An accelerated salt fog type test (ASTM B-117) may be used to evaluate this phenomenon which is characteristic for certain types of coatings.[29]

The National Association of Corrosion Engineers has established standard test TM-01-74 for the laboratory evaluation of protective coatings used as linings in immersion service, as well as tests for one-side exposure of coated panels.

Several general references relative to the testing of organic coatings are listed at the end of the chapter.[30–38] Inasmuch as with the exception of filiform corrosion, the resistance of organic coatings to corrosives is a function of the polymer used in the coating, much of the testing information in Chapter 7 is also applicable to coatings.

REFERENCES

1. J. A. Capp, *ASTM Proc.* **14**(2), 274 (1914).
2. A. Becerra and R. Darby, *Corrosion* **30**(5), 153 (1974).
3. L. L. LaQue, *Materials and Methods* **35**(2), 77 (1952).
4. Anon., *ASTM Proc.* **15**(1), 220 (1905).
5. W. C. Porter, "Painting Steel," Report BMS-102, *Building Materials and Structures,* Washington, DC, (1944).

6. R. Caryl and C. Rheineck, *Off. Dig. J. Paint Technol. Eng.* **37**(2), 481 (1965).
7. C. E. Hoey and H. A. Hipwood, *JOCCA* **57**(5), 151 (1974).
8. W. R. Tooke, *Transp. Res. Rev.* **692,** 14 (1978).
9. Robert H. Perry and Cecil H. Chilton, Eds., Section 23 *Chemical Engineer's Handbook,* 5th ed., McGraw-Hill, New York, 1973.
10. O. B. J. Fraser, D. E. Ackerman, and J. W. Sands, *Ind. Eng. Chem.* **19,** 332 (1927).
11. P. L. Gordon and R. Gordon, *Paint and Varnish Manual,* Wiley-Interscience, New York, 1955.
12. T. Touchsaen and A. Leidheiser, *Corrosion* **28**(11), 435 (1972).
13. R. Desalme, *J. Steel* (London), **164**(2), 240 (1950).
14. R. C. Bacon, J. T. Smith, and F. M . Rugg, *Ind. Eng. Chem.* **40,** 161 (1948).
15. M. M. Goldberg and N. D. Tomoshov, *Zavodskaya Lab.* **15,** 951 (1949).
16. R. N. O'Brien and H. Kolny, *Corrosion* **8,** 263, 354 (1978).
17. H. Strange, K. Bauer, and A. Bitldorf, *Plaste Kautsch* **26**(8), 465 (1979).
18. E. Clarke, J. R. Carlin, and W. E. Barboor, *Elec. Eng.* **70,** 35 (1951).
19. S. Moser and R. K. Wilt, *Ind. Eng. Chem.* **41,** 2334 (1949).
20. C. J. H. Monk, *J. Oil Colour Assoc.* **35,** 178 (1952).
21. I. N. Finhorn and J. D. Crecca, *Mater. Prot.* **2**(7), 10 (1963).
22. A. E. Schuh and E. W. Kern, *Ind. Eng. Chem.* (*Anal. Ed.*) **3,** 72 (1931).
23. P. Whitely, *JOCCA* **32**(3), 410 (1969).
24. E. S. Matsui, *J. Paint Technol.* **41**(553), 455 (1967).
25. J. Rybicky and J. P. Morton, *J. Paint Technol.* **47**(605), 57 (1975).
26. C. F. Sharman, *Nature* **20,** 621 (1944).
27. J. A. Boylon, *J. Paint Technol.* **44**(555), 48 (1972).
28. E. L. Koehler, *Corrosion* **33**(6), 209 (1977).
29. A. C. Holler, *Off. Dig. J. Paint Technol. Eng.* **35**(466), 1163 (1963).
30. R. F. Eisenberg, R. R. Kraybill, and R. B. Seymour, Chapter 1 in *Laboratory Engineering and Manipulation,* E. S. Perry and A. Weissberger, Eds., Wiley-Interscience, New York, 1979.
31. H. W. Chatfield, *The Science of Surface Coating,* D. Van Nostrand, New York, 1962.
32. N. I. Gaynes, *Testing Organic Coatings,* Noyes Data Corp., Park Ridge, NJ, 1977.
33. R. R. Meyers and J. S. Long, *Treatise on Coatings,* Marcel Dekker, New York, 1969–79.
34. P. Nylen and E. Sunderland, *Modern Surface Coatings,* Wiley-Interscience, New York, 1965.
35. H. F. Payne, *Organic Coating Technology,* John Wiley & Sons, New York, 1954.
36. R. B. Seymour, *Hot Organic Coatings,* Reinhold, New York, 1960.
37. R. B. Seymour, "Chemical Resistance of Nonmetallic Materials," in *Analytical Chemistry in Industry,* I. M. Kolthoff, P. J. Elving, and F. H. Stross, Eds., Wiley-Interscience, New York, 1976.
38. D. H. Solomon, *The Chemistry of Organic Film Formers,* John Wiley & Sons, New York, 1967.

CHAPTER 7

Testing of Plastics for Chemical Resistance

The chemical reactions for the deterioration of self-supporting plastics are essentially the same as those for polymeric coatings. However, tests for the former are independent of a metallic substrate and their rapid degradation may result in catastrophic failure. Likewise, there is no telltale appearance of degradation or rusting with self-supporting plastics and therefore more objective tests for the prediction of failure are essential.

TESTS FOR THERMAL DEGRADATION

Although the failure of plastics in corrosive environments is the result of chemical degradation (chemidegradation) rather than thermal degradation, primary covalent bonds are cleaved in both processes, therefore the magnitude of bond energies [E (cal/mol)] and energies of activation [E_a (cal/-mol)] are important. The temperature at which a bond loses half its weight when heated for 30 minutes ($T_{1/2}$) has been used as a criterion of thermal stability.[1] Accordingly, expression (7.1) has been developed for this first-order reaction.

$$T_{1/2} = 5.75\,E + 185 \tag{7.1}$$

The $T_{1/2}$ value in kelvin temperatures for several typical polymers are as follows:

Polytetrafluoroethylene (PTFE)	509
LDPE	404
PP	387
Polychlorotrifluoroethylene (PCTFE)	380
Polystyrene (PS)	364

Polyphenylene oxide (PPO)	345
Polymethyl methacrylate (PMMA)	327
PVC	260

Presumably, $T_{1/2}$ values are significant for thermal degradation but activation energies E_a must also be considered when chemical degradation occurs at high $T_{1/2}$ values. As stated in Chapter 4, the deterioration of polymers involves swelling and cleavage of covalent bonds by oxidation or hydrolytic attack, as well as thermal degradation.[2]

Of course, there may be abnormal surface, color, dimensional, hardness, and percent weight changes (PWC) but only the latter approaches the basis for objective testing. However, whereas changes in slope for PWC versus time curves may be significant, they are not always reliable unless accompanied by more objective test data.[3]

CHANGE IN WEIGHT TESTS

The American Society for Testing Materials (ASTM) has established a test in which PWC is reported after seven days immersion in 51 chemical reagents (ASTM D-543).[4–7] Change in weight measurements has been correlated with other significant changes in monitoring the chemidegradation of rigid plastics[8] and changes of ±2% in the weight of polymers are concluded acceptable.[9]

WATER ABSORPTION

The water absorption of certain plastics such as polyfluorocarbons and polyacetals is very low and has negligible effect on the properties of the plastic. However, more polar plastics such as nylons and cellulose acetate may absorb significant amounts of water. In the standard test (ASTM D-570), the increase in weight of a test specimen is measured after 24-hr immersion in water at 25°C.

PERMEATION TESTS

High absorbency and permeation of corrosives into a plastic may contribute to catastrophic failure. As might be anticipated, the rate of permeation may be monitored by measuring changes in the dielectric constant of the plastic specimen.[10] The extent of penetration of the corrosive liquid may be determined by embedding probes at different depths[11] or by skinning (i.e., removing thin layers of the plastic).[12] The latter technique permits the reporting of data as mm/yr as in the corrosion testing of metals. Permeation testing is essential for evaluating the chemical resistance of fiberglass-reinforced plastics (FRP).

CHANGES IN HARDNESS

Changes in hardness after immersion in a corrosive solution may be measured by using a Rockwell hardness tester (ASTM D-785) in which the depth of indentation of a steel ball is measured. The change in penetration of blunt needles has been used to show the chemical degradation of asphalt.[13] The change in hardness of elastomers (ASTM D-865) and a modified Fras brittle point test have been used to monitor the chemical degradation of other polymers.[14] The change in hardness of reinforced plastics (FRP) is monitored by indentation hardness using a Barcol Impressor Tester in ASTM method D-2583.

TENSILE STRENGTH

The change in tensile or pulling strength of a standard dogbone-shaped test specimen (ASTM D-638) has been widely used as a test for chemical degradation of thermoplastics,[15] and the percent retention of tensile strength has been recommended for the evaluation of the chemical resistance of phenolic laminates.[16]

The term threshold aging period has been coined to indicate the extrapolated time during which the tensile properties of polyurethane decrease catastrophically under conditions of high humidity.[17] Prot stress rupture tests have been used to predict the chemical resistance of FRP.[18] This endurance test is based on loaded test samples with moderate loads, such as 65 percent of total tensile strength.

Polymers have characteristic shift factors when tested under loads over long periods of time in specific environments. For example, when a value of 1 is assigned for the shift factor of a stressed polymer immersed in hexane, the values in isopropyl alcohol are as follows: PVC, 0.08; polycarbonate (PC), 3; ABS, 120; and styrene–acrylonitrile copolymer, 1500.[19,20]

STRESS RELAXATION TESTS

Tests on stress versus time on standard dogbone test specimens in specific environments compared to those in air have been used to determine the usefulness of plastics in corrosive environments.[21] Creep and stress relaxation test data have been used to predict the performance of PVC in corrosive environments.[22]

STRESS RUPTURE TESTS

The stress required for 60-sec rupture of a tensile test specimen immersed in specific solvents has also been used for the evaluation of the usefulness of plastics in specific organic environments.[23]

IMPACT STRENGTH

The percent decrease in Izod impact strength (ASTM D-256) has been used to evaluate the chemical degradation of phenolic and polyester laminates.[24,25,26] The percent retention of both impact and flexural strength has been used to measure the resistance of FRP pipe to corrosives.[27] The depth of weathering attack on PVC has been correlated with the loss in impact strength.[28]

COMPRESSIVE STRENGTH

Changes in the compressive strength of a test specimen after exposure to a corrosive environment have been used to determine the usefulness of plastic concrete (ASTM C-267, D-695).[29] The chemical resistance of epoxy[30] and polyester[31] plastics has been estimated by use of percent compressive strength retention values.

FLEXURAL STRENGTH AND MODULUS

In the absence of fissures or other defects, changes in flexural or bending strength or modulus (ASTM D-419, C-581)[32] of a test specimen or thermoset plastic equipment, have been considered as useful tests for chemical degradation.[33] Several investigators believe that chemical resistant polyester plastics will undergo an initial loss of flexural strength when exposed to corrosive environments but the values will level off at a constant strength level.[34,35] Thermal cycling[36] and thermal mechanical analytical tests[37] have also been proposed.

INSTRUMENTAL CHARACTERIZATION

Chemical degradation is a chemical reaction in which the rate of attack may be measured by monitoring the concentration of new chemical groups in the polymer. These reactions are both time and temperature dependent, thus the rate of degradation of a loaded specimen will increase exponentially as the temperature is increased in accordance with the Arrhenius equation.[38]

Torsion pendulum tests (ASTM D-2236) in which an initial torsional displacement is applied and the sample is allowed to decay freely, as well as dynamic mechanical testing, have been used to measure the rate of chemi-degradation.[39] Although these tests are usually made on a molded specimen, many instrumental tests such as attenuated total reflectance spectroscopy (ATR) may be used as nondestructive tests.

Extremely small samples may be removed from the surface of a polymer and tested by infrared spectroscopy (IR), UV spectroscopy, nuclear resonance spectroscopy (NMR), pyrolysis gas chromatography (PGC), and differential thermal analysis (DTA).

IR spectroscopy has been used to predict the useful life of HDPE,[40] and to study solvent absorption,[41] oxidation,[42] and other degradative reactions of polymers.[43] UV spectroscopy has been used to identify carbonyl groups in degraded polymers[44] and NMR spectroscopy has been used to study the aging of rubber.[45]

PGC has been used to characterize the degradative products of epoxy resins[46] and DTA has been used to monitor the degradation of cellulose derivatives[47] and the effect of citric acid on polyacetals.[48]

ATR spectroscopic measurements which may be made on plastic equipment provide data similar to IR spectroscopy. ATR spectroscopy has been used to monitor the rate of increase of oxidized sites during the weathering of PVC.[49]

Additional information on testing may be found in the references at the end of this chapter.[50–56]

REFERENCES

1. V. V. Koshak and S. V. Vinogradova, *Russian Chem. Ser.* **37,** 11 (1968).
2. R. A. Cass and O. H. Fenner, *Ind. Eng. Chem.* **56**(8), (1964).
3. A. E. Lever and J. Rhys, *The Properties and Testing of Plastics*, Chemical Publishing Co., New York, 1958.
4. F. A. Champion, *Corrosion Testing Procedures*, Chapman Hall, London, 1952.
5. O. H. Fenner, in *Managing Corrosion Problems with Plastics*, Vol. III, National Association of Corrosion Engineers, Houston, TX, 1977, p. 287.

6. E. Evans, *Plastics as Corrosion-Resistant Materials*, Pergamon Press, Oxford, England, 1966.
7. R. B. Seymour and R. H. Steiner, Chapter 6 in *Plastics for Corrosion Resistant Applications*, Reinhold Publishing Co., New York, 1955.
8. L. S. Van Delinder, *Mat. Prot.* **2**(3), 30 (1963).
9. A. O. Fisher, *Tools and Techniques for Laboratory Corrosion Testing*, NACE, Houston, TX, 1962.
10. J. Delmonte and E. Sama, Proc. 18th SPE Ann. Techn. Manage. Conf. Reinf. Plast. Div., 1963.
11. R. A. Cass and O. H. Fenner, *Ind. Eng. Chem.* **56**(8) (1964).
12. R. A. Cass and O. H. Fenner, *Mat. Prot.* **4**(10) (1965).
13. L. R. Sheppard, *Corrosion* **17,** 1577 (1961).
14. R. L. Berger, *SPEJ* **20,** 630 (1954).
15. H. Schnell, *Inst. Eng. Chem.* **51,** 157 (1959).
16. J. Delmonte, *Mod. Plast.* **20**(5), 91 (1942).
17. S. Steingeiser and W. C. Darr, *Rubber Chem. Technol.* **37,** 38 (1964).
18. E. J. Kerle, *Chem. Eng. Progr.* **61**(4), 74 (1965).
19. G. E. Fulmer, *Polymer Eng. Sci.* 280 (1967).
20. Y. Snegawa, H. Hojo, T. Ideda, and Y. Okamura, *Materials and Research Standard* **3,** 550 (1965).
21. S. Okuda, in *Managing Corrosion with Plastics*, National Association of Corrosion Engineers, Houston, TX, 1977, p. 70.
22. L. Martha, *Maunyag Gumi* **16**(3), 80 (1979). [*Chem. Abstracts* **91,** 57, 960 (1979).]
23. L. M. Robeson, in *Managing Corrosion with Plastics*, National Association of Corrosion Engineers, Houston, TX, 1977, p. 166.
24. C. J. Straka, *Mod. Plast.* **20**(11), 80 (1943).
25. P. M. Goldfarb, PB 161912, Off. Tech. Ser., Washington, DC, 1959.
26. T. S. McQuarrie and R. D. Gould, Proc. 35th Ann. SPI Conf. Reinf. Plast. Comp. Inst., 1980.
27. R. Hoehn and W. Heling, *Reinf. Plast.* **4**(3), 12 (1963).
28. M. S. Stangl, K. Binder, and H. Tschauler, *Chem. Kunstst Aktuell* **31** (1977).
29. R. W. Childers, *Mat. Prot.* **2**(3) 18 (1963).
30. H. L. Jones and A. N. Ciancianulo, *Chem. Eng. Prog.* **61**(4), 65 (1965).
31. P. M. Goldfarb, PB 16912, Off. of Techn. Ser. Washington, DC, Dec. 1959.
32. H. S. Loveless, in *Testing of Plastics*, Chapter 10, Vol. 2, J. V. Schmitz, Ed. Interscience, New York, 1966.
33. S. A. Moore and A. Weber, *Chem. Eng.* **72**(20), 176 1963.
34. R. A. Cass and O. H. Fenner, *Mat. Prot.* **4**(10) 1965.
35. R. C. Allen, Proc. 35th Ann. SPI Conf. Reinf. Plast. Comp. Inst., 1980.
36. T. S. McQuarrie and R. D. Gould, Proc. 35th Ann. SPI Conf. Reinf. Plast. Comp. Inst., 1980.
37. R. D. Gould, T. S. McQuarrie, and K. Abrycki, Proc. 35th Ann. SPI Conf. Reinf. Plast. Comp. Inst., 1980.
38. R. R. Lantos, *Chem. Tech.* **7,** 184 (1977).

39. C. E. Pratt, *Plast. Eng.* **34**(8), 28 (1978).
40. S. Orbais, E. Bonai, and H. Korskina, *Chem. Kunstst Aktuell* **43** (1977).
41. R. S. Foster and V. B. Volkening, *Corrosion* **15**(10), 85 (1959).
42. R. B. Seymour, H. S. Tsang, and D. Warren, *Polym. Eng. Sci.* **7**(1), 55 (1967).
43. H. H. G. Jellinek, G. F. Flausman, and F. J. Kryman, *J. Appl. Polym. Sci.* **13,** 107 (1969).
44. V. E. Gray and J. R. Wright, *Polym. Eng. Sci.* **5**(4), 284 (1965).
45. H. Y. Chen and N. E. Lewis, *Anal. Chem.* **36**(7), 1394 (1964).
46. J. M. Stuart and D. S. Smith, *J. Appl. Polym. Sci.* **2,** 2195 (1963).
47. D. F. Arsenau, *Can. J. Chem.* **39,** 1916 (1961).
48. DES Cagliostro, S. Riccitello, and J. A. Parker, *J. Macromol. Sci. Chem. A* **3,** 1601 (1969).
49. G. C. Marks and G. Butler, *J. Macromol. Sci. Chem. A* **12,** 569 (1978).
50. H. E. Atkinson, *Managing Corrosion with Plastics*, National Association of Corrosion Engineers, Houston, TX 1977, p. 223.
51. H. R. Edwards, *Reinforced Plastics Comp. Inst. Tech. Rev.*, Society of the Plastics Industry, New York, 1976, Sect. 6E.
52. O. H. Fenner, Chapter 4 in *Handbook of Plastics and Elastomers*, C. A. Harper, Ed. McGraw-Hill, New York, 1975.
53. J. Frados, Chapter 30, *SPI Plastics Engineering Handbook,* Van Nostrand Reinhold Co., New York, 1976.
54. J. S. McDermott, *Managing Corrosion with Plastics*, National Association of Corrosion Engineers, Houston, TX, 1977, p. 185.
55. R. B. Seymour, Section D-1 in *Analytical Chemistry in Industry*, Vol. 3, I. M. Kolthoff, P. J. Elvering, and F. H. Stross, Eds., Wiley-Interscience New York, 1976.

CHAPTER 8

Protective Coatings

The use of plastic structures in corrosive environments is relatively new but surface coatings have been used for the protection of various substrates for thousands of years. It is of interest to note that Noah used bitumens or slime as a surface coating for the wood in his ark and the early Egyptians used molten beeswax as a weather resistant coating.[1]

The art gradually advanced to the use of polymerizable (drying) oils such as linseed oil with a minimum amount of solvent,[2] to high solvent coatings in the 1940's–1970's. Because evaporating solvents contribute to atmospheric pollution, the protective coatings industry has now returned to the high solids-low solvent products.[3]

Whereas a protective coating is usually adhered to the substrate, the same polymers used for surface protection may be used independently as films. Inasmuch as the emphasis in this book is on chemical resistance of polymers, both adhered and free films will be considered in this chapter.

The first successful plastic film was cellulose nitrate made by the evaporation of a mixture of ethanol and ethyl ether from collodion. Regenerated cellulose or cellophane and cellulose acetate were also pioneer films.

Both cellulose nitrate and cellulose acetate were used widely as films in the first part of the twentieth century. Dope, used as a coating on fabric aircraft fusilage and wings during World War I, was a solution of cellulose acetate in acetone. Duco lacquer used to protect the first automobiles from the weather was a pigmented solution of cellulose nitrate. Although these films and coatings were serviceable, they have been replaced to a large extent by synthetic polymers.

OIL BASED COATINGS

Until recently, paints based on linseed oil (from flax seed) and white lead were used almost exclusively for the protection of wood surfaces. Some of these formulations included natural resins. These were subsequently replaced by oil-soluble phenolic resins.

Other polymerizable oils used in the paint industry included the following: Oiticica oil from a Brazilian tree (*Licania rigida*), dehydrated castor oil (DCO), perilla oil (Japan), conophor oil (Nigeria), safflower oil, stillingia oil (Chinese tallow tree), tall oil (pine oil), menhaden and other fish oils, soybean oil, cottonseed oil, and tung oil (*Aleurites fordi*).

All of these "drying oils" are glyceryl esters of unsaturated acids such as licanic acid (4-keto-octadeca-9,11,13-trienoic acid) found in oiticica oil and shown in (8.1).

$$CH_3(CH_2)_3CH{=}CHCH{=}CHCH{=}CH(CH_2)_4CO(CH_2)_2COOH \quad (8.1)$$

It is believed that, in the presence of driers or siccatives which are soluble heavy metal salts (e.g., those of lead, manganese, or cobalt) oxygen is inserted between the hydrogen atom and the alpha carbon of the unsaturated oil molecule to produce a peroxide-free radical. As illustrated in (8.2), the double bond remains intact.[4]

Formation of a peroxide-free radical in the "drying" of a paint film:

$$\text{—}CH_2CH{=}CH\text{—}CH_2\text{—} + O_2 \xrightarrow{Pb^{2+}} CH_2\text{—}CH{=}CH\text{—}\underset{\underset{\underset{\dot{O}}{|}}{\overset{|}{O}}}{CH}\text{—} \quad (8.2)$$

Subsequent reactions (8.3) in the "drying process" are typical free radical chain polymerization reactions that are also typical for the synthesis of ethylenic polymers such as LDPE.

Propagation and termination reactions in free radical chain polymerization:

$$\begin{aligned} ROO\cdot + RH &\longrightarrow ROOH + R\cdot \\ R\cdot + R\cdot &\longrightarrow R{:}R \end{aligned} \quad (8.3)$$

Inasmuch as the "drying oils" are usually polyunsaturated esters, the chain extension reaction (polymerization) occurs in more than one position in the molecule and a three-dimensional insoluble network polymer is produced. In the absence of film defects, these coatings are useful for the protection of wooden structures against weathering. However, the chemical resistance of oil-base paints is limited and they should not be recommended for use in aggressive corrosive environments.[5]

NATURAL RESINS

Shellac, one of the few resins of animal origin, is an excretion from an insect (*Coccus lacca*) which feeds on twigs of trees in southern Asia. This ethanol-soluble resin consists of esters, lactones, and anhydrides of carboxylic acids including aleuritic acid (9,10,16-trihydroxypalmitic acid). Shellac coatings are resistant to aliphatic hydrocarbons and have been widely used as wood coatings. This resin has little use as a corrosion resistant coating and has been displaced to a large extent by synthetic resins.

Amber, an old fossil resin found on the shores of the Baltic sea; dammar, a more recent fossil resin found in the East Indies; sandarac, a fossil resin from northwest Africa; and mastic, a fossil resin from Greece, have been used as lacquers and varnishes. However, these colorless resins, like shellac, have been replaced to a large extent by synthetic resins.

BITUMINOUS MATERIALS

Bituminous coatings have been used for over two thousand years, and are still used as surface coatings.[6,7,8] However, whereas the original coatings were based on naturally occurring bitumens, most of today's bituminous coatings are based on coal tar pitch and petroleum still residues. These dark-film-forming products are usually filled with clay or other fillers. The annual production of asphalt, coal tar, and natural bitumens in the United States is 27, 3.4, and 0.2 million tons, respectively. Solutions of asphalt in aliphatic hydrocarbon solvents, called cut-back, are used as protective coatings on pipe and steel structures. Filled asphaltic compositions called mastics are used as heavy coatings on steel vessels.[9] Of course, extremely large quantities of filled asphalt are used for roofing and road construction.

CELLULOSICS

Cellulose composites such as wood have been used in corrosive environments for centuries, but cellulose derivatives were unknown prior to 1845 when Schönbein nitrated cellulose.[10] He mistakenly assumed the derivative to be a nitro compound.

That this inorganic ester was resistant to sulfuric and nitric acids was demonstrated by the presence of residues of these acids in the classical cellulose nitrate. Wilson and Green patented solutions of cellulose nitrate (CN) in 1884, and these lacquers are still in use today.

Cellulose nitrate (nitrocellulose) lacquer made from esters with a low degree of polymerization (DP) has been used as a protective coating on metals, primarily on automobiles and furniture, since 1923.[1] Cellulose nitrate has a glass transition temperature (T_g) of 53°C and is usually plasticized by alkyl adipates. Cellulose nitrate also is used as an explosive and obviously is flammable.

Cellulose acetate (CA), produced by Schutzenberger in 1865[12] and by Franchimont in 1874[13] by the acetylation of cellulose with acetic anhydride in the presence of acetic and sulfuric acids, has a degree of substitution (DS) of about three. This so-called triacetate has a relatively low solubility parameter (9*H*) and is soluble in chlorinated alkane solvents.

This triacetate is used to a limited extent as a coating. However, pipe extruded from this cellulose ester has been widely used for transporting natural gas.

The principal cellulose ester is the more polar secondary ester obtained by the partial saponification of cellulose triacetate. Lacquers produced by dissolving the partially saponified cellulose acetate in acetone are widely used as a film and to a small extent as a coating. The dope used for aircraft prior to the 1920's and fingernail polish are cellulose acetate lacquers. The film is widely used in reverse osmosis applications.[14]

Cellulose acetate butyrate (CAB) has been used for almost a half century as an inner liner for safety glass and as a coating for cloth and wood. Appropriately compounded CAB signs have been used satisfactorily outdoors for many years. Cellulose esters of organic acids are flammable but much less so than CN.

Commercial methylcellulose has a low DP. Since it is usually water soluble, it has few applications as a coating for use in corrosive environments. Ethylcellulose (EA) is a colorless polymer used both as a solution type and as a melt coating. EA is less resistant to acids than CA but is much more resistant to alkalies.

RUBBER DERIVATIVES

Because of its tackiness, natural rubber (NR, *Hevea braziliensis*) has limited use as a coating. However, its derivatives, which are nontacky thermoplastics, have been used widely as coatings and films. The classical rubber derivative, chlorinated rubber (Parlon), is the oldest and most widely used rubber derivative.[15]

These derivatives are available with different degrees of substitution (DS) and are insoluble in aliphatic hydrocarbons but soluble in aromatic hydrocarbon solvents. The derivative with 67 percent chlorine is still used as a chemical resistant coating for concrete such as that in swimming pools.

Cyclized rubber (Pliolite), prepared by the isomerization of NR with Lewis acids such as chlorostannic acid, has also been widely used as a protective coating. Cyclized rubber is soluble in aliphatic hydrocarbon solvents and is more resistant to heat than chlorinated rubber.

Cyclized rubber has been used as an adhesive for NR tank linings and as a chemical resistant coating. Although the chemical resistance of cyclized rubber is very good, it has been displaced by other coatings, such as styrene–butadiene copolymers because of its higher costs.

Oxidized rubber (Rubbone) has also been used as a protective coating. Rubber hydrochloride (Pliofilm), which also has very good resistance to corrosives, has been widely used as a packaging film. However, because of high costs, most of these rubber derivatives have been replaced by less expensive synthetic polymers.

ALKYD RESINS

The word "alkyd" is a euphonious expression coined by Kienle for use in place of the term alcid.[16] The latter indicates that these resins are condensation products of *al*cohols and a*cids*.

Although the term was used originally for oil-modified glyceryl phthalate resins, it usually applies to resinous products obtained by the condensation of polyhydric alcohols and polycarboxylic acids. The alkyds are produced at an annual rate of 225,000 tons and are the most widely used coating resins.

The first polyester resin was reported by Berzelius in 1847[17] but the first useful resin of this type was a glyceryl phthalate prepared by Watson Smith in 1901.[18] These esters, including those in which part of the phthalic anhydride is replaced by the saturated and unsaturated aliphatic monocarboxylic acids, were studied intensively by many scientists at General Electric Company prior to the investigations of Kienle and Hovey in 1930.[19]

Air-drying alkyds are widely used as building and maintenance coatings. The baked alkyds are used as automotive and appliance coatings. Inasmuch as alkyds are esters, they have limited resistance to alkalies but have fair resistance to aqueous acids and excellent resistance to aqueous salt solutions.

The chemical resistance of allylic resins, which are produced by the polymerization of diallyl phthalate, is similar to that of alkyds. However, allylic resins have limited use as coatings.

Although over 550,000 tons of unsaturated polyesters are used annually in the United States, most of this resin is used as a fiberglass-reinforced plastic.[20] Solutions of condensation products of ethylene glycol, phthalic anhydride, and maleic anhydride in styrene are used to a limited extent as *in situ* polymerizable coatings. However, the high concentration of styrene vapors released during the coating process of these solvent-free coatings restricts the use of these resins as coatings. Yet, filled unsaturated polyesters are used as plastic concrete.

Solutions of vinyl esters, which are esters of bisphenol-A and methacrylic acid, are used as coatings. These, like other polyesters, have excellent resistance to aqueous salt solutions, fair resistance to aqueous acid solutions, and poor resistance to alkalies.

Alkyds, with water soluble groups, may be used as water-borne coatings. Although the chemical resistance of these resins is inferior to that of regular alkyds, they are being used for the protection of aircraft and metals.[21]

PHENOLICS

Phenolic resins were introduced by Leo Baekeland in the early 1900's. Varnishes consisting of solutions of condensation products of p-phenylphenol and formaldehyde have good resistance to acids and salt solutions. Their resistance to solvents is improved when they are blended with alkyds.

Novolak resins produced by the condensation of formaldehyde with an excess of phenol under acidic conditions have been used to a limited extent as coatings. Resole prepolymer resins produced by the condensation of stoichiometrical amounts of formaldehyde and phenol under alkaline conditions have been used as heat reactive coatings.

Filled solvent-free resole resins are used as plastic concrete, and pigmented solutions of these reactive resins are used as baked coatings for drilling pipe, printing plate backing, and as drum and tank linings.[22] Cured phenolic coatings have a high crosslink density and have excellent resistance to nonoxidizing acids and salts, good resistance to organic solvents, and fair resistance to alkalies. Strong alkaline solutions tend to soften phenolic coatings but since this is an attack on a pendant hydroxyl group, it does not involve cleavage of the polymer chains.

EPOXY RESIN COATINGS

Epoxy resins, produced by the condensation of bisphenol-A and epichlorohydrin, are produced at an annual rate of 155,000 tons in the United States. Approximately 25,000 and 15,000 tons of epoxy resins are used as binders for fiberglass-reinforced plastics and tooling, respectively, and the bulk of the annual production is used for bonding and adhesives (8,000 tons), flooring (8,000 tons), and protective coatings (70,000 tons). The product mix for the latter is as follows: appliance finishes (4,000 tons), automotive primers (11,000 tons), can and drum coatings (16,300 tons), pipe coatings (4,500 tons), maintenance (13,000 tons), and trade sales coatings (20,000 tons).

Epoxy resin prepolymers contain epoxy terminal groups and hydroxyl groups on all other repeat units, and may be cured by reaction with polyamines which react with the epoxy groups at room temperature or by reaction with cyclic anhydrides at elevated temperatures. They may also be cured by UV radiation.[23]

These resins, patented by Schlack in Germany in 1934 and by Castan in Switzerland in 1940,[24] are particularly useful for chemical resistant coatings. They are characterized by excellent adhesion to the substrate, excellent resistance to alkalies and salts, good resistance to nonoxidizing acids, and fair resistance to organic solvents.

POLYURETHANES

Polyurethanes (developed by O. Bayer in Germany in the late 1930's) are also characterized by excellent adhesion to all types of substrates[25] as well as excellent resistance to abrasion. Over 865,000 tons of polyurethane resin (PU) are used annually in the United States but most of this PU is used as flexible (608,000 tons) and rigid foam (258,000 tons).

The American Society for Testing Materials (ASTM) has grouped PU coatings into five categories, namely: Type 1, alkyd oil-modified PU; Type 2, isocyanate-terminated PU coatings which cure by contact with atmospheric moisture; Type 3, blocked PU which must be heat-cured to release the blocking agent; Type 4, two-package system consisting of isocyanate-terminated resin and catalyst; and Type 5, two-package system consisting of a polyol and an isocyanate-terminated adduct.

The two-component PU systems have very good resistance to corrosive environments and organic solvents. Based on accelerated hydrolytic stability tests, it has been concluded that PU liner pads will retain their integrity when immersed in water at 35°C for 24 years.[26]

AMINO RESINS

The term "amino resins" is used to describe those produced by the condensation of formaldehyde with urea or melamine. Almost 600,000 tons of amino resins are produced annually in the United States. Less than 50,000 tons are used as protective coatings, but over 400,000 tons of these resins are used as bonding and adhesive compositions.

Amino resin coatings may be used as two-component systems in which an acid serves as the curing agent. Butylated amino resins, which are soluble in hydrocarbon solvents, are also used as coatings. Both urea and melamine resins are used as crosslinking agents in alkyds but amino resin coatings have limited use in corrosive environments.[27]

SILICONES

Siloxane polymers have a backbone of alternating silicon and oxygen atoms and were incorrectly called silicones by Kipping, who assumed that these synthetic resins were ketones. These resins were commercialized by Dow Corning and General Electric Company in the late 1930's.[28]

Silicone resins, which are produced by the hydrolysis of chloro or alkoxy silanes, have excellent resistance to elevated temperatures and are water repellant. Silicone coatings are widely used as heat resistant coatings and as they are transparent to UV radiation, they have excellent resistance to weathering.

VINYL CHLORIDE POLYMERS

Polyvinyl chloride, described by Regnault in 1838 and by Baumann in 1872, is one of the oldest synthetic thermoplastics. However, since it is not readily dissolved by solvents, "vinyl coatings" based on the more soluble copolymer of vinyl chloride and vinyl acetate were not commercialized until the 1930's.[29]

Over 2,800,000 tons of vinyl chloride polymers are produced annually in the United States, but only about 50,000 tons of these resins are used as protective coatings. The copolymers of vinyl chloride and blends of PVC with acrylic resins or chlorinated polyethylene are more flexible than PVC itself. However, it is customary to add liquid plasticizers such as dioctyl phthalate (DOP) to vinyl chloride resins to assure greater flexibility.

There has been considerable concern about the toxicity of residual vinyl chloride monomer (VCM) in PVC. However, vinyl coatings are essentially free from VCM. Unless polar groups such as anhydride, carboxyl, or hydroxyl groups are present in the vinyl chloride copolymer, it will not adhere well to the substrate. Therefore, it is customary to use a primer before the application of a vinyl coating. Vinylite (VAGH), which contains some hydroxyl groups, has good adhesion to metal.

Vinyl polymers are soluble in ketone solvents and it is customary to use a blend of ketones and aromatic hydrocarbons as the solvent for vinyl coatings. Vinyl coatings may also be applied as emulsions, powder coatings, and as dispersions of PVC in organic solvents (organosols), liquid plasticizers (plastisols), or water [aqueous powder suspension (APS)].

PVC coatings have excellent resistance to aqueous solutions of acids, alkalies, salts, and oils. Copolymers of vinyl chloride and vinyl acetate (Vinylite) and plasticized PVC are not as resistant to corrosive environments as PVC itself. PVC coatings tend to degrade at elevated temperatures. However, commercial vinyl coatings contain stabilizers to assure maximum performance at elevated temperatures.

FLUOROCARBON COATINGS

Polytetrafluoroethylene (PTFE, Teflon), which was synthesized accidentally and patented by Plunkett in 1938, is one of the most chemically resistant polymers.[30] However, since it is insoluble in solvents and does not adhere readily to substrates, it is not used as a solution coating. Polychlorotrifluoroethylene (CTFE, Kel F) and polyhexafluoropropylene are slightly

Figure 8.1. Fluorocarbon insulated wire used in Montreal Metro Subway System. (Courtesy of Allied Chemical Corp.)

Figure 8.2. Fluorocarbon coated diaphragm valve. (Courtesy of Allied Chemical Corp.)

more flexible and almost equal in chemical resistance to PTFE, but they are not soluble enough to be used as solution coatings.

However, copolymers of chlorotrifluoroethylene and hexafluoropropylene with vinylidine fluoride (VDF) are useful as chemical resistant gaskets but are not used as solution coatings.

Polyvinylidine fluoride (PVF_2) is not as resistant to elevated temperatures as PTFE but it is soluble in polar solvents such as dimethyl sulfoxide (DMSO) and dimethylformamide (DMF). Coatings of PVF_2 are not resistant to fuming sulfuric acid but are resistant to most acids, alkalies, salts, and nonpolar solvents.

Vinyl fluoride was synthesized in 1901 but was not polymerized until 1934. The polymer (PVF, Tedlar) is soluble in hot polar solvents such as DMF. PVF films are resistant to most acids, alkalies, salts, and nonpolar solvents. Wire and valves coated with a polyfluorocarbon soluble coating are shown in Figures 8.1 and 8.2.

ACRYLICS

Polyacrylic acid has been known for over 125 years although the esters of this acid were not commercialized until the twentieth century. Otto Rohm promoted the use of acrylic coatings and molding compounds. In the 1950's acrylic resins were accepted as automotive coatings. Over 265,000 tons of acrylic polymers are produced annually in the United States.

Acrylic coatings may be applied as solutions, aqueous emulsions, and powder. High-solids acrylic coatings may be curved by UV radiation. The crosslinked (cured) coatings are more resistant to elevated temperatures than the linear polymer. Acrylic coatings are characterized by excellent resistance to UV radiation and weathering. They have good resistance to aqueous solutions of nonoxidyzing acids, salts, and alkalies. The methacrylates are more resistant to alkalies than the acrylates.[31]

STYRENE POLYMERS

In addition to the 2,000,000 tons of SBR, over 2,300,000 tons of styrene polymers are produced annually in the United States. However, very little of these polymers are used as coatings. Emulsions of copolymers of styrene and butadiene have been used to a small extent as coatings. Styrene was polymerized by Simon in 1839 and the polymer (PS) was commercialized in the 1930's.[32]

Styrene, of course, is the polymerizable solvent in unsaturated polyesters. Oriented PS film and ABS sheet and pipe are used in corrosive environments. Polystyrene is soluble in aromatic hydrocarbons but is resistant to aqueous solutions of salts and alkalies and has moderate resistance to nonoxidizing acids.

POLYOLEFINS

Polyolefins are produced at an annual rate of 7,500,000 tons in the United States and are the most widely use polymers. Inasmuch as the homopolymers, namely, LDPE, HDPE, and PP, are crystalline, they are insoluble in solvents below their melting points. However, they are applied as coatings through fluidized bed and electrostatic spray techiques.[3]

Polyolefin wire coatings are applied by extrusion techniques. Since chlorosulfonated polyethylene (Hypalon) is soluble in organic solvents, it is used as a solution coating. This derivative, as well as the polyolefins, is resistant to aqueous solutions of salts, alkalies, and acids.

POLYAMIDES

Nylon-66, produced by Carothers by the condensation of hexamethylenediamine and adipic acid in the 1930's, and other nylons including nylon-6, are useful as abrasion resistant coatings.[34] Because of the limited solubility of these polyamides, the coating is usually applied over a primer by the fluidized bed technique. These polyamides have good resistance to aqueous solutions of salts and alkalies.

Polyamides, produced by the condensation of ethylenediamine and a dibasic acid such as a dimerized fatty acid, are used as hot melt and water-dispersed coatings.[35] These polyamides are also used as curing agents for epoxy resins. The polyamide coatings have good resistance to weather and aqueous salt solutions.

ENGINEERING POLYMERS

The use of polymers, such as the polyacetals, polyphenylene oxide, polybutylene terephthalate, polysulfones, and polyphenylene sulfides as coatings has been limited because they are not readily soluble. However, they have good chemical resistance and are used as powder and melt coatings. Polyphenylene sulfide (Ryton) has been used as a chemical resistance melt coating for pipe, including the pipe threads.[36]

APPLICATION TECHNIQUES

The surface of the object to be protected by a protective coating should be adequately prepared. Wood and concrete surfaces should be brush cleaned and all rust and contaminants removed from steel surfaces. Shot or sand-blasted surfaces are highly desirable and the primer or first coat should be applied as soon as possible to the cleaned surface.

Solution coatings may be applied by brushing, dipping, flow coating, curtain coating, roll coating, or spraying.[37] Air spraying, which is commonly used, is fast but inefficient. Hydrostatic spraying at high pressures results in less overspray and electrostatic spraying in which the atomized coating is attracted to the surface of the metal is almost 100 percent efficient.

Powdered polymers have been applied to relatively small objects by the fluidized bed technique. In this technique, air is used to force the powdered polymer into the heated surface of the object, which is in the upper section of a closed tank. The coated object is removed from the fluidized bed tank and heated in an oven to assure a continuous coating.

Water-borne coatings may be applied by brushing, flow coating, or spraying techniques. Objects as large as an automobile chassis are coated by the electrodeposition of resin from an aqueous system. Inasmuch as the resin is deposited by high voltage (200 V) onto the oppositely charged object, a uniform coating is produced. After removal of solvents, thermoset coatings may be cured by heating or by use of electron beam (EB) or ultraviolet radiation.

High solids compositions, plastisols, and plastic mortars are discussed in Chapter 9. Additional information on coatings may be found in the references at the end of this chapter.[38–45]

REFERENCES

1. A. H. Sabin, *The Industrial and Artistic Technology of Paint and Varnish,* John Wiley & Sons, New York, 1906.
2. C. A. Smith, *London Journal of Arts* (1825).
3. R. B. Seymour, *Australian Oil and Colour Chemists Association,* **14**(11), 4 (1977).
4. W. O. Lundberg, Ed., *Autoxidation and Autioxidants,* Vol. I, Interscience, New York, 1961.
5. H. A. Gardner and G. G. Sward, *Physical and Chemical Examination of Paint Varnishes, Lacquers, Colors,* 12th ed., Gardner Lab., MD, 1962.
6. H. Abraham, *Asphalts and Allied Substances,* 6th ed., Van Nostrand Co., Princeton, NJ, 1963.
7. A. J. Hoiberg, *Bituminous Materials; Asphalt, Tars and Pitches,* Interscience, New York, 1964.
8. R. B. Seymour, *Hot Organic Coatings,* Rheinhold, New York, 1960.
9. S. H. Alexander, Chapter 45 in *Applied Polymer Science,* J. K. Craver and R. W. Tess, Eds., Org. Coat. and Plast. Chem. Div., American Chemical Society, Washington, DC, 1975.
10. H. Skolnik and L. F. McBurney, *Tappi* **37**(9), 1902 (1954).
11. R. F. Wint and C. W. Vanderslice, Chapter 57 in *Applied Polymer Science,* J. K. Craver and R. W. Tess, Eds., Org. Coat. and Plast. Chem. Div., American Chemical Society, Washington, DC, 1975.
12. P. Schutzenberger, *Compt. Rend.* **61,** 485 (1865).
13. A. Franchimont, *Compt. Rend.* **89,** 711 (1879).
14. J. D. Crowley, Chapter 56 in *Applied Polymer Science,* J. K. Craver and R. W. Tess, Eds., Org. Coat. and Plast. Chem. Div., American Chemical Society, Washington, DC, 1975.
15. L. Bateman, *The Chemistry and Physics of Rubber-Like Substances,* John Wiley & Sons, New York, 1963.
16. R. H. Kienle, *Ind. Eng. Chem.,* **41,** 726 (1949).
17. J. Berzelius, *Rappt. Ann. Inst. Geol. Congrie.* 26 (1847).
18. W. Smith, *J. Soc. Chem. Ind.,* **20,** 1073 (1901).
19. H. J. Lanson, Chapter 37 in *Applied Polymer Science,* J. K. Craver and R. W. Tess, Eds., Org. Coat. and Plast. Chem. Div., American Chemical Society, Washington, DC, 1975.

20. H. V. Boenig, *Unsaturated Polyesters,* Elsevier Publishers, New York, 1962.
21. D. Campbell, *Am. Paint and Coat. J.* **63**(33), 58 (1979).
22. A. Knopp and W. Scheib, *Chemistry and Application of Phenolic Resins,* Springer-Verlag, New York, 1979.
23. W. R. Watt, Chapter 1 in *Epoxy Resin Chemistry,* 116 ACS Symposium Series, R. S. Bauer, Ed., Washington, DC, 1974.
24. H. Lee and K. Neville, *Handbook of Epoxy Resins,* McGraw-Hill, New York, 1967.
25. K. C. Frisch, Chapter 54 in *Applied Polymer Science,* J. K. Craver and R. W. Tess, Eds., Org. Coat. and Plast. Chem. Div., American Chemical Society, Washington, DC, 1975.
26. M. A. Mendelsohn and G. B. Rosenblatt, Chapter 11 in *Durability of Macromolecular Materials,* R. R. Eby, Ed., No. 95, ACS Symposium Series, Washington, DC, 1979.
27. C. P. Vale and W. G. K. Taylor, *Aminoplastics,* Illiffe Books Ltd., London, 1964.
28. E. G. Rochow, *An Introduction to the Chemistry of Silicones,* John Wiley & Sons, New York, 1951.
29. G. O. Curme and S. D. Douglas, *Ind. Eng. Chem.* **28,** 1123 (1934).
30. R. J. Plunkett, U.S. Patent 2,330,654 (1961).
31. W. H. Brendley, G. V. Calder, and L. A. Wetzel, Chapter 55 in *Applied Polymer Science,* J. K. Craver and R. W. Tess, Eds., Org. Coat. and Plast. Chem. Div., American Chemical Society, Washington, DC, 1975.
32. R. H. Boundy, R. F. Boyer, and S. M. Stoesser, *Styrene, Its Polymers, Copolymers and Derivatives,* Reinhold, New York, 1952.
33. R. A. V. Raff and J. B. Allison, *Polyethylene,* Wiley-Interscience, New York, 1956.
34. M. I. Kohan, *Nylon Plastics,* John Wiley & Sons, New York, 1973.
35. T. Bradley and W. Johnson, *Ind. Eng. Chem.* **16,** 90 (1944).
36. H. W. Hill, Chapter 13 in *Durability of Macromolecular Materials,* R. K. Eby, Ed., ACS Symposium Series, Washington, DC, 1979.
37. E. P. Miller, Chapter 21 in *Applied Polymer Science,* J. K. Craver and R. W. Tess, Eds., Org. Coat. and Plast. Chem. Div., American Chemical Society, Washington, DC, 1975.
38. J. H. Briston, *Plastic Films,* John Wiley & Sons, New York, 1974.
39. C. A. Harper, Ed., *Handbook of Plastics and Elastomers,* McGraw-Hill, New York, 1975.
40. R. R. Meyers and J. S. Long, *Treatise On Coatings,* Marcel Dekker, New York, 1972.
41. P. Nylen and E. Sunderland, *Modern Surface Coatings,* Interscience, New York, 1965.
42. H. F. Payne, *Organic Coating Technology,* John Wiley & Sons, New York, 1954.
43. R. B. Seymour, *Hot Organic Coatings,* Reinhold, New York, 1960.
44. D. H. Solomon, *The Chemistry of Organic Film Formers,* John Wiley & Sons, New York, 1967.
45. O. J. Sweeting, *The Science and Technology of Polymer Films,* Wiley-Interscience, New York, 1971.

CHAPTER 9

High Solids Compositions, Plastisols, Pastes, Powders, and Caulking Compositions

As stated in Chapter 8, it has been necessary to reduce the solvent content of many solution coatings in order to comply with EPA regulations. This may be accomplished by the use of reactive oligomers (low-molecular-weight polymers) and higher application temperatures.[1]

HIGH SOLIDS COMPOSITIONS

These systems include the reactions of oligomers with hydroxyl and isocyanate terminal groups, dibutyltin dilaurate (DBTDL) or stannous octoate (SnOCT) catalysts, acrylics with hydroxyl groups and methylolmelamine, polyester oligomers with hydroxyl end groups and methylolmelamine, and epoxy prepolymers and polyamines.

The high solids polyurethanes (PU) have been used for solvent-free coatings[2] but the trend is toward the production of large articles by use of reaction injection molding (RIM) techniques. In the RIM method, the polyol with catalysts and the diisocyanate oligomer are fed into a molding press through separate inlets and the polymerization is allowed to take place in the mold.

Hydroxyl functional acrylic oligomers such as polyhydroxylethylmethacrylate (HEMA) may be cured in the presence of a methylolmelamine and an acid catalyst.[3] Oligomers of polyesters with terminal hydroxyl groups may be cured in much the same way be use of a methylolmelamine. The resistance to solvents and corrosives of these high solids resins is dependent on the relative amounts of reactants used.[4]

The cure of epoxy prepolymers may be effected by use of ditertiary amines (Dabco) or mercapto esters such as pentaerythritol tetrakis (mercaptopropionate, mercaptate Q 43 ester). New formulations of epoxy resins have good weatherability.[5] Cycloaliphatic epoxides, with formulas such as shown in (9.1), have been used in heat reactive (100°C) two-component "matched pair" polycarboxylic acid systems in the presence of stannous octoate catalysts and silicone anti-cratering agents.[6] Small amounts of triethylamine may be added to the prepolymer mixture to extend the potlife.

A cycloaliphatic epoxide:

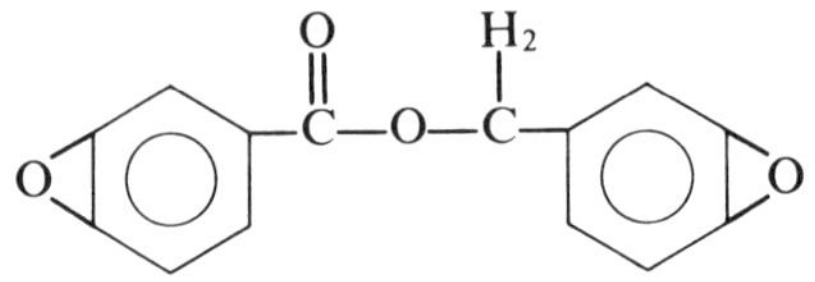

(9.1)

PLASTISOLS

A plastisol is a viscous suspension of fine particles (75–200 μm) of PVC in a liquid plasticizer.[7] This paste is converted irreversibly to a flexible solid when heated at 175–190°C. Although PVC is the most widely used polymer, copolymers of vinyl chloride and vinyl acetate or acrylic acid have been used. The latter have better adhesion to the substrate.[8]

The most widely used plasticizer for PVC plastisols is di-2-ethylhexyl phthalate (DOP) but small amounts of other plasticizers are usually added to achieve faster fusion and better physical properties of the plasticized PVC. Stabilizers, pigments, and fillers may also be present.

Fumed silica or bentonite clays have been used as thickeners in vinyl plastisols and volatile organic liquids are present in organosols. The latter may be prepared by the addition of hydrocarbon diluents and volatile ketones (dispersants) to the plastisols.[9]

The plastisol compositions containing large amounts of aminated bentonite clay, metallic soaps, or filmed silica are called plastigels.[10] Those made with aminated bentonite clay may be shaped at ordinary temperatures and fused in that shape without the use of molds.

Crosslinking agents may be added to plastisols so that the fused product is thermoset. Plastisols are used in relatively large quantities for the production of coatings, pipe joints, PVC foams, and castings, as well as for rotational and slush molding of PVC articles. The chemical resistance of these products is comparable to that of PVC with similar compositions made by other fabrication techniques.

POLYMER PASTES

Finely divided polymers such as polymethyl methacrylate (PMMA), bitumens, cellulosics, and cumarone resins may be dispersed in a liquid monomer which may be polymerized to produce a solid plastic. Filled two-component PMMA-methyl methacrylate systems have been used for the manufacture of dentures.[11]

Solid compositions have also been produced by cooling hot mixtures of filled cellulose esters and solid plasticizers. Mastics and pipe joints have been prepared by dispersing asphalt and fillers in a small amount of solvent.

POLYMER POWDERS

As stated in Chapter 8, articles may be coated by polymer powders using fluidized bed or dispersion techniques. These methods, as well as electrostatic spray and rotational molding, may be used with a large variety of polymers for the production of coatings, films, and plastic articles.

Polyethylene, polypropylene, cellulose acetate butyrate, epoxies, polyesters, acrylics, and PVC are all available as powders which may be used for powder or electrostatic coatings, or rotational molding. Plastic parts may be produced by spraying a powdered polyolefin in an electrostatic field into a heated open mold. The mold is cooled by cold water and the solid part is removed.[12]

Electrostatic spray techniques have been used to apply thermosetting polyester powders to metal objects such as fences and farm equipment.[13] Electrostatic-sprayed polyurethane coatings have been applied and cured at 65°C and used as protective coatings in corrosive environments.[14]

In rotational molding (rotomolding), a polymer powder is heated in a closed cavity or mold which is rotated biaxially in a heated chamber. Cellulosics, nylon, polyacetals, polycarbonates, polyurethane, polyesters, polypropylene, and polystyrene powders, as well as PVC powder and plastisols, have been molded by this process.[15]

CAULKING COMPOSITIONS

The original caulking compositions of bitumens or vegetable oil-based materials have been replaced to some extent by compositions based on polybutene, butyl rubber, natural and synthetic elastomers, polysulfides, polyurethanes, and silicones.[16] Polybutene, produced by the cationic polymerization of isobutylene, is a tacky polymer that may be filled and used as a caulking material.

Because of its ready availability and curability, butyl rubber, a copolymer of isobutylene (95) and isoprene (5), is often used in place of polybutene. Both caulking compositions are resistant to high humidity including steam, and have excellent weatherability. These polymers are resistant to aqueous solutions of monoxidizing acids, alkalies, and salts.

Depolymerized rubber, neoprene, SBR, and polymers of hydroxyl- and carboxyl-terminated butadiene have chemical resistant properties similar to those of polybutene. The hydroxyl terminated products may be used as homopolymers (poly BD) or copolymers with styrene or acrylonitrile. It is customary to react the hydroxyl-terminated polymers with diisocyanates to produce polyurethanes.

The carboxylated butadienes may be used as homopolymers or as copolymers with acrylonitrile (Hycar, CTBN). It is customary to cure these polymers with epoxy resins using polyamines such as triethylenetetramine as curing agents. Another related caulking composition is made from butadiene polymers with vinyl pendant groups (Ricon).

The use of the original polysulfide polymers as caulking compositions is limited. These oil resistant elastomers were produced by the condensation of sodium polysulfide and ethylene dichloride.[17] However, curable liquid polymers (LP-2) that have been obtained by the reduction of these solid polysulfide elastomers are widely used as caulking compositions. These liquids have mercapto end groups and are readily crosslinked by lead peroxide, manganese dioxide with morpholine, 2,4-toluyl diisocyanate with *N*-methyl-2-pyrollidone, or epoxy resins.[18]

These compositions may be filled with carbon black, calcium carbonate, or clay and have good resistance to aqueous solutions of nonoxidizing acids, alkalies, and salts. They are resistant to aliphatic hydrocarbons but are swollen in benzene and ketone solvents.

Polyurethane caulking compositions and sealants are usually two-component systems consisting of a hydroxyl-terminated and an isocyanate-terminated prepolymer. One-package polyurethane caulking composition may also be cured by exposure to moisture in the air.

The first silicone caulking compositions (RTV) were two-component systems consisting of an acid-terminated polysiloxane and a polysilicate.[19] The most widely used silicone caulking composition consists of a hydroxyl-terminated polydimethylsiloxane, a filler such as silica, and a reactive silane crosslinking compound. The later are trifunctional silanes such as trimethoxymethylsilane.[20] These compositions have excellent weatherability and retain their physical properties over a wide temperature range.

D. B. Dahm, who estimates construction as a $160,000,000,000 industry, has classified the caulking and sealants used by this industry into the following types[21]: (1) nondrying systems, (2) solvent or solution, (3) drying oil or oxidative type, (4) emulsion or water-based systems, and (5) chemical-reaction systems. He cites the following as examples of products illustrative of these types: (1) PVC rope (tape) caulk, (2) butyl caulk, (3) blown soybean oil caulk, (4) PVC latex caulk, and (5) moisture cure and two-component urethanes, moisture cure, and two-component silicones, two-component epoxies, and polysulfides.

Additional information on these products may be found in the references at the end of this chapter.[22-28]

REFERENCES

1. J. A. Antonelli, *Amer. Paint and Coat. J.* **59**(42), 44 (1975).
2. L. C. Curtis, Proceedings of USM 4th Water-Borne and High Solids Coatings Symposium, New Orleans, Feb., 1977.
3. R. R. Kuhn, Proceedings of USM 6th Water-Borne and High Solids Coatings Symposium, New Orleans, 1979.
4. S. N. Belote, Proceedings of USM 6th Water-Borne and High Solids Symposium, New Orleans, 1979.
5. R. S. Bauer, Proceedings of USM 6th Water-Borne and High Solids Coatings Symposium, New Orleans, 1979.
6. O. W. Smith, J. V. Kolski, and J. G. Kucsma, Proceedings of USM 4th Water-Borne and High Solids Coatings Symposium, New Orleans, Feb., 1977.
7. W. L. Semon, U.S. Patent 1,929,453 (1933).
8. D. W. Ward, *SPE J.* **28**(5), 44 (1972).
9. G. M. Powell, *Modern Plastics* **28**(10), 129 (1951).
10. R. W. Quarles, E. T. Severs, A. C. Frechtling, and H. S. Carpenter, *Modern Plastics* **29**(5), 99 (1952).
11. J. A. Hetherington, *Plastics. Inst. Trans.* **17**(29), 25 (1945).
12. K. J. Chivers, Chapter 17 in *SPI Plastics Engineering Handbook*, J. Frados, Ed., Van Nostrand-Reinhold, New York, 1976.
13. G. Adcock, *Mod. Paint and Coat. J.* **68**(9), 54 (1978).
14. W. Brushwell, *Am. Paint and Coat. J.* **62**(51), **48**(52), **58**(55), (1978).
15. D. J. Ramozzoli, Chapter 14 in *SPI Plastics Engineering Handbook*, J. Frados, Ed., Van Nostrand-Reinhold, New York, 1975.
16. R. B. Seymour, Chapter 8 in *Plastics, Mortars, Sealants and Caulking Compounds*, R. B. Seymour Ed., ACS Symposium Series #13, Washington D.C., 1979.
17. S. M. Martin and J. C. Patrick, *Ind. Eng. Chem.* **28,** 1145 (1936).
18. M. A. Schulman and A. D. Yazujian, Chapter 11 in *Plastic Mortars, Sealants, and Caulking Compounds*, R. B. Seymour, Ed., ACS Symposium Series #113, Washington, DC, 1976.

19. J. F. Hyde, U.S. Patent 3,571,039 (1951).

20. J. M. Klosowski and G. A. L. Gant, Chapter 10 in *Plastic Mortars, Sealants, and Caulking Compounds,* R. B. Seymour, Ed., ACS Symposium Series #113, Washington, DC, 1979.

21. D. B. Dahm, Proceedings of USM 5th Water-Borne and High Solids Coatings Symposium, New Orleans, Feb. 1978, p. 38.

22. L. S. Buchoff, Chapter 8 in *Handbook of Plastics and Elastomers,* C. H. Harper, Ed., McGraw-Hill, New York, 1975.

23. C. A. Clark, T. J. Kraus, and D. M. Ward, "Vinyl Dispersions," Chapter 16 in *SPI Plastics Engineering Handbook,* J. Frados, Ed., Van Nostrand-Reinhold, New York, 1976.

24. E. M. Fettes and J. S. Jorczak, "Polysulfide Polymers," Chapter 11 in *Polymer Processes,* C. E. Schildknecht, Ed., Interscience, New York, 1956.

25. H. A. Sarvetnick, Ed., *Plastisols and Organosols,* Van Nostrand-Reinhold, New York, 1972.

26. R. B. Seymour, *Plastic Mortars, Sealants and Caulking Compounds,* ACS Symposium Series #113, Washington, DC, 1979.

27. R. B. Seymour, *Modern Plastics Technology.* Reston Publishing, Reston, MD, 1975.

28. J. J. Sokol and R. C. Hendrickson, "Powder Coatings" Chapter 17 in *SPI Plastics Engineering Handbook,* J. Frados, Ed., Van Nostrand-Reinhold Co., New York, 1976.

29. W. D. Todd, "Paste Techniques," Chapter 14 in *Polymer Processes,* C. E. Schildknecht, Ed., Interscience, New York, 1956.

CHAPTER 10

Foams, Castings, and Plastic Mortars

Although cellular plastics are not usually considered as materials of construction for use in corrosive environments, they are used for insulating pipes and vessels. Inasmuch as rigid foams are stronger on a weight basis than solid plastics, they also will be used more as materials of construction.

FOAMS

Natural rubber sponges and cushions were produced over a half century ago by using a chemical blowing agent such as carbon dixide from the decomposition of sodium carbonate and by whipping air into creamed rubber latex before vulcanization. The use of the latter latex foams continues but the major growth has been in polyurethane (PU) foams and expanded polystyrene (EPS). The annual United States production of flexible and rigid PU foams is 608,000 and 258,000 tons, respectively. Over 400,000 tons of EPS are also produced annually.

Urea formaldehyde (UF) foams were patented in Germany in the early 1930's.[1] These cellular products were prepared by frothing an aqueous solution of UF prepolymer with high agitation and adding an acid to polymerize the frothed resin.

Phenolic foams were made inadvertently by Leo Baekeland as a result of the evolution of gaseous products during the molding of phenolic plastics (PF).[2] Commercial PF foams, called Troporit p and Thermazote were produced in Germany and England during World War II. PF foams are produced by the addition of an acid curing agent to a mixture of resole PF, surfactant, and blowing agent.

Epoxy resin (EP) foams have been produced by adding a blowing agent such as a volatile fluorocarbon to a mixture of EP prepolymer, surfactant, and amine curing agent such as diethylenetriamine. However, most EP foams are syntactic foams produced by the addition of hollow glass beads (microballoons) to EP resins.[3]

Polystyrene foam may be produced by the extrusion of a mixture of a volatile liquid such as pentane and PS[4] or by molding expandable PS beads.[5] The present production is divided about equally between the two types of EPS.

PU foams were produced by Bayer in the late 1930's by the addition of water to a mixture of a diol and a diisocyanate.[6] The carbon dioxide propellant was produced by the decomposition of the unstable carbamic acid produced from the reaction of water and the isocyanate.

The *in-situ*-formed carbon dioxide has been replaced by other propellants. Flexible PU foams are produced by the reaction of a flexible polyether with terminal hydroxyl groups and a diisocyanate. The rigid PU foams used for insulation are produced by the reaction of a rigid polyol and a diisocyanate. These cellular plastics may be produced as slabs or foamed in place.

Other plastic foams have been produced from silicones,[7] PVC,[8] polyolefins,[9] polyesters, polyfluorocarbons, cellulose acetate, polyimides, and polysulfones.[10] The chemical resistance of foams is identical to that of the solid polymers. However, hot solvents may cause a collapse of the foam structure. Additional information on foams may be found in a book on the subject.[11]

CASTINGS

The first phenolic and urea formaldehyde resins were castings produced by the addition of an acid catalyst to the prepolymers. These thermoset cast plastics were augmented by the commercialization of cast thermoplastic acrylic plastics by Rohm in the 1920's, polyesters in the 1940's, and nylon-6, epoxies, and polyurethanes in the 1950's.

Phenolic and epoxy resin castings are readily produced by adding catalysts, an acid or polyamine, respectively, to liquid prepolymers. The complete polymerization may be assured by post heating of the castings. Solution of unsaturated polyesters in styrene may be polymerized by the addition of an initiator such as benzoyl peroxide and a promoter such as dimethylaniline.[12]

Precautions must be taken to prevent high concentrations of styrene monomer in the atmosphere near the polyester casting process. The shrinkage may be decreased by substituting vinyltoluene for part of the styrene. The heat of polymerization (exotherm) in large castings may be reduced by the addition of a small amount of alpha-methylstyrene to the prepolymer.

Castings of polyurethane are produced by mixing a diisocyanate such as tolyl diisocyanate (TDI) or methylene bis(4-phenyl isocyanate) (MDI) with a polyol such as polyether or polyester with terminal hydroxyl groups. Organo-tin compounds are used as catalysts and silicone copolymers are sometimes used as surfactants with these reactants. These castings are produced on a large scale in the reaction injection molding process (RIM).

Large castings of nylon-6 are readily produced by the anionic polymerization of caprolactam. Nylon-6 castings contain unreacted monomer but because of their high molecular weight they are as strong as molded nylon-6 products.

Methyl methacrylate is widely used for the casting of PMMA sheet, rod, and other shapes.[13] It is customary to polymerize part of the monomer to a syrup by heating with a peroxy initiator before pouring the solution of PMMA in monomer into a mold and continuing the heating process. A slurry of finely divided PMMA in monomer may also be used.

This casting process may be carried out on a continuous basis. Acrylic castings under the name of "reactive adhesives" or "honeymoon adhesives" are widely used as coatings.[14]

The liquid resin molding process (LRM) may also use diallyl phthalate (Dapon), diethylene glycol bis(allylcarbonate) (CR-39), triallyl cyanurate (TAC), and diallyl chlorendate (DAC). The latter yields flame resistant castings.[15]

PLASTIC MORTARS

When the castings discussed in this chapter are polymerized in the presence of fillers, they are called plastic mortars or plastic concrete. Although these composites were developed originally for joining brick and tile for use in chemical resistant structures, they are now widely used in construction in place of hydraulic cements. The availability of automatic mixing equipment has aided the increase in use of these traditional plastic composites.[16]

Silicate mortars have been used as joining cements for acid resistant brick in acid tanks for more than a half century.[17] These proprietary products are made by mixing a filler containing sodium fluosilicate (Na_2SiF_6) with an aqueous solution of sodium silicate. The latter should have a specific gravity of about 38°Baumé (Be) and a ratio of Na_2O to SiO_2 of 4 to 13.

The filler usually consists of a graded silica aggregate. The hardening of the mixture is based on the hydrofluoric acid produced by the hydrolysis of the sodium fluosilicate. The setting time is inversely proportional to the concentration of the sodium fluosilicate and the temperature.

The compressive strength of a typical silicate cement is 10,000 psi

(68,950 kPa). This product, like all plastic mortars, has a high coefficient of expansion, in this case, 2.0 cm/cm/°C $\times 10^{-5}$. Silicate cements are characterized by high water absorption (10%). These products are resistant to all acids except hydrofluoric acid, and all solvents and all salts except fluorides and alkaline salts at temperatures up to 400°C. Silicate cements should not be used with plain water or aqueous alkaline solutions.

Sulfur cements are mixtures of sulfur and graded silica or carbon fillers used as hot melts. They may be plasticized by polyethylene sulfide. Since sulfur undergoes a transition from rhombic to monoclinic crystalline form at 95°C, sulfur cements must not be used above that temperature.

A silica-filled sulfur cement will have a compressive strength of 8000 psi (55,160 kPa) and a coefficient of expansion of 3.0 cm/cm/°C $\times 10^{-5}$. Sulfur cements are resistant to nonoxidizing inorganic acids and aqueous solutions of nonalkaline salts at temperatures up to 90°C. Carbon-filled sulfur cements are resistant to hydrofluoric acid, but neither silica nor carbon-filled sulfur cements should be used with nonpolar organic solvents.

The original resinous mortar was a silica-filled phenolic cement called Asplit. This pioneer product consisted of a phenolic resole resin which was mixed—on the job site—with a silica or carbon filler containing a latent acid setting agent such as benzene sulfonyl chloride or p-toluene sulfonic acid.[18]

Carbon-filled phenolic resin cements have a compressive strength of 9000 psi (62,055 kPa), a coefficient of linear expansion of 6.3 cm/cm/°C $\times 10^{-5}$, and a low water absorption of 0.2%. The carbon-filled phenolic resin cements are resistant to nonoxidizing acids, nonalkaline salts, and many organic solvents at temperatures up to 175°C. Like other phenolic plastics, they are not satisfactory for use in strong alkaline solutions.

Furan-resinous cements based on polymers of furfuryl alcohol were developed in the 1930's as a replacement for phenolic resin cements in alkaline environments.[19] The liquid resin in the original product called Alkor was simply a solution of polybutyl methacrylate in furfuryl alcohol. The furfuryl alcohol was polymerized by the acid (p-toluene sulfonic acid) present in the carbon filler.

Inasmuch as the liquid component of modern furan cements is a polymer of furfuryl alcohol, the exothermic reaction is more readily controlled.[20] Furan cements, like phenolic cements, are used to join brick and tile used in corrosive environments.

The compressive strength of a typical furan cement is 8750 psi (60,331 kPa). Its coefficient of expansion, water absorption, and temperature limitations are similar to those listed for phenolic cements. Furan cements are not as resistant to oxidizing acids as phenolic cements but they are resistant to aqueous alkaline solutions.

The art used for curing epoxy resins with polyamines has been adapted by the producers of chemical resistant cements.[21] These cements are more ductile than phenolic or furan cements and may be used for the construction of monolithic floor coatings as well as for the joining of brick and tile.[22]

Epoxy resin cements have a compressive strength of 10,000 psi (68,950 kPa), water absorption of 0.3%, a maximum service temperature of 90°C and a coefficient of linear expansion of 4.4 cm/cm/°C $\times 10^{-5}$. These values will vary with the ratio of resin to filler and the curing agent used. Epoxy resin cements are resistant to nonoxidizing acids, salts, alkalies, and polar organic solvents.

Since none of the available resinous cements were resistant to oxidative chemicals such as chlorine dioxide, polyester resin cements under the trade name of Vitroplast were developed in the early 1950's for use in the pulp and paper industry.[23]

The liquid resin binder consists of a solution of an unsaturated polyester and an N,N-dimethylaniline promoter in styrene. The filler is mixed with the liquid resin—on the jobsite—and consists of a graded silica aggregate and a peroxy initiator such as benzoyl peroxide.

These polyester resins have a compressive strength of 10,000 psi (68,950 kPa), a coefficient of expansion of 3.2 cm/cm/°C $\times 10^{-5}$, and a water absorption of 0.2%. These cements are resistant to inorganic acids and salts at temperatures up to 80°C. Although these cements are not recommended for use in strong alkaline environments, their resistance to alkalies can be improved by using more alkaline resistant resins.

Precautions must be taken to reduce the concentration of styrene monomer in the atmosphere when polyester cements are being applied. Other resinous cements may be modified to meet specific chemical resistance requirements. Also, polyurethanes, nylon-6, and other casting resins may be filled and used as plastic cements.

REFERENCES

1. I. G. Farbenindustrie, German Patent 636,658 (1933).
2. L. H. Baekeland, *Ind. Eng. Chem.* **1,** 149 (1909).
3. J. Delmonte, Proceedings 15th Ann. Conf. SPI Reinf. Plast. Div., Chicago, Feb. 1960.
4. C. G. Munters and J. G. Tandberg, U.S. Patent 2,023,204 (1935).
5. J. J. P. Staudinger, B. K. Kelly, and A. Cooper, British Patent 605,863 (1948).
6. O. Bayer, *Angew. Chem.* 459, 257 (1947).
7. H. L. Vincent, *Plastic Foams,* K. C. Frisch and J. H. Saunders, Eds., Marcel Dekker, New York, 1973.

8. A. C. Werner, *Plastic Foams,* K. C. Frisch and J. H. Saunders, Eds., Marcel Dekker, New York, 1973.
9. D. J. Sundquist, *Plastic Foams,* K. C. Frisch and J. H. Saunders, Eds., Marcel Dekker, New York, 1973.
10. K. C. Frisch, *Plastic Foams,* K. C. Frisch and J. H. Saunders, Eds., Marcel Dekker, New York, 1973.
11. K. C. Frisch and J. H. Saunders, Eds., *Plastic Foams,* Marcel Dekker, New York, 1973.
12. V. L. Montani in *SPI Plastics Engineering Handbook,* J. Frados, Ed., Van Nostrand-Reinhold Co., New York, 1976.
13. J. F. Woodman, *SPI Plastics Engineering Handbook,* J. Frados, Ed., Van Nostrand-Reinhold Co., New York, 1976.
14. J. W. Prane, Chapter 9 in *Plastic Mortars, Sealants, and Caulking Compounds,* R. B. Seymour, Ed., ACS Symposium Series #113, Washington, DC, 1979.
15. R. B. Seymour and R. H. Steiner, Chapter 9 in *Plastics for Corrosion Resistant Applications,* Reinhold Publishing Co., New York, 1955.
16. R. B. Seymour, Proceedings of 5th SPE PACTEC, Los Angeles, 1980.
17. J. G. Vail, *Soluble Silicates,* Reinhold Publishing Co., New York, 1952.
18. R. B. Seymour, Chapter 1 in *Plastic Mortars, Sealants, and Caulking Compounds,* R. B. Seymour, Ed., ACS Symposium Series #113, Washington, DC, 1979.
19. C. R. Payne and R. B. Seymour, U.S. Patent 2,266,049 (1944).
20. R. H. Leitheiser, M. E. Londrigan, and C. A. Rude, Chapter 2 in *Plastic Mortars, Sealants, and Caulking Compounds,* R. B. Seymour, Ed., ACS Symposium Series #113, Washington, DC, 1979.
21. J. M. Sosa, Chapter 3, in *Plastic Mortars, Sealants, and Caulking Compounds,* R. B. Seymour, Ed., ACS Symposium Series #113, Washington, DC, 1979.
22. P. Maslow, Chapter 4 in *Plastic Mortars, Sealants, and Caulking Compounds,* R. B. Seymour, Ed., ACS Symposium Series #113, Washington, DC, 1979.
23. R. B. Seymour, Chapter 5 in *Plastic Mortars, Sealants, and Caulking Compounds,* R. B. Seymour, Ed., ACS Symposium Series #113, Washington, DC, 1979.

CHAPTER 11

Polymeric Linings and Thermoplastic Pipe and Structures

Polymeric linings are actually heavy coatings that are usually applied in sheet form and adhered to the internal surfaces of a steel vessel. The pioneer sheet lining was vulcanized natural rubber and this lining is still in use today.

POLYMERIC LININGS

Linings are distinguished from protective coatings by an arbitrary definition based on thickness. The coatings are usually less than 2.5 mm thick whereas the linings are usually over 3.0 mm thick. Inasmuch as the emphasis is on thickness, heavy coatings such as mastics may be designated as linings.

However, most linings used in the chemical and metal processing industries are sheet membrane linings which form an impervious thick barrier between the corrosive and the surface of the vessel. Linings may be applied to almost any surface but their major use is for the protection of steel tanks against corrosion by aqueous acidic solutions.[1]

The sheet linings are adhered to the surface of the supporting structure by an adhesive such as cyclized rubber and care is taken to make all seams impervious to liquids. Unprotected thermoplastic linings such as plasticized PVC and vulcanized natural rubber linings may be used at temperatures up to 65°C.

However, when protected by a brick sheathing, joined by plastic mortars, these linings may be used at higher temperatures. The maximum operating temperature of an uninsulated brick-lined steel tank is dependent on the thickness of the brick sheathing. Of course, the temperature of the organic membrane lining itself must not be above 65°C. The temperature drop resulting from 10 and 20 cm of brick is 35 and 50°C, respectively.[2]

The most widely used membrane linings are asphalt, natural rubber, neoprene, plasticized PVC, SBR, and butyl rubber. Bituminous coatings such as asphalt are limited to the room temperature for storage of aqueous solutions.

Corrosives at slightly higher temperatures may be used with bituminous linings containing crosslinking agents such as epoxy resins. When these linings on concrete tanks are protected by brick sheathings, these tanks may be used for the storage of hot corrosive liquids. These linings are resistant to aqueous solutions of nonoxidizing acids, alkalies, salts, and polar solvents such as ethanol. They are not resistant to nonpolar solvents such as oils.

Soft natural rubber or SBR sheet linings are adhered by a solvent-softened cyclized rubber primer applied on both the tank and rubber surface. It is customary to use a roller to prevent any air entrapment and to use warm sheets to insure pliability.

Natural rubber sheets are overlapped and the heat-softened edges are stitched with a mechanical serrated roller. The lining in the tank is then vulcanized by heating in a large oven.

Soft natural rubber linings are resistant to aqueous solutions of nonoxidizing acid such as dilute sulfuric acid, alkalies, and salts. The surface of these linings is hardened by aqueous solutions of hydrochloric acid. Hard rubber (ebonite) is used to a limited extent and a triple-ply lining with a layer of ebonite has been used as well.[3]

Natural rubber linings have also been applied by spraying thickened compounded rubber latex. Since these linings are compounded with ultra-accelerators, they are vulcanized at ordinary temperatures. Neoprene linings have also been applied from solution and as sheets.

Inasmuch as neoprene is pervious to hydrochloric acid, it is customary to include a thin layer of natural rubber hardened by hydrochloric acid. Neoprene has better resistance than natural rubber to oils and hydrocarbon solvents.[4]

The resistance of SBR rubber linings is similar to that of natural rubber. Linings of polyisobutylene tend to cold flow but cured butyl rubber, chlorobutyl rubber, and ethylene–propylene copolymers (EDPM) are serviceable at temperatures up to 60°C. These linings are softened by nonpolar solvents and oils.

CHLOROSULFONATED POLYETHYLENE

Hypalon linings applied from solution or by sheet have chemical resistance similar to EPDM. The chemical resistance of sheet linings of copolymers of vinylidene chloride and acrylonitrile (Saran) is similar to that of natural rubber at temperatures up to 40°C.

Plasticized PVC (Koroseal) sheet linings are more resistant to oxidizing acids such as chromic and nitric acids than natural rubber. These linings may be used at temperatures up to 40°C and at somewhat higher temperatures when a dual lining of plasticized and rigid PVC is used. The temperature resistance of chlorinated PVC (PVDC) is superior to that of PVC.

PVC and PVDC (CPVC) linings have also been protected by fiberglass-reinforced plastics (GRP).[5] Linings of polyvinylidene fluoride (Kynar) and other polyfluorocarbons (Teflon FEP) may be adhered by impressing a knitted fiberglass fabric into the plastic and adhering this backing to the metal surface (Armalon).[6] These linings are resistant to almost all acids, alkalies, salts, and solvents at temperatures as high as those that can be withstood by the adhesive used.

Epoxy, polyester, vinyl ester, and polyurethane trowelling compositions have also been used satisfactorily as heavy coatings. These, as well as the sheet linings, have also been used to line metal pipe.[7,8,9]

THERMOPLASTIC PIPE

Most thermoplastics have also been extruded as pipe[10] and molded as valves and fittings. Coal-tar-impregnated paper tubing has been used for transporting domestic wastes but is not recommended for use with corrosive liquids.

Polyvinylidene chloride pipe (Saran) was extruded and used to a limited extent prior to World War II. This pipe is useful for service with nonoxidizing acids and salts at temperatures up to 50°C.[11] The techniques developed for this pipe were adapted to rigid PVC pipe in Germany during World War II. Although polyvinylidene chloride is still used as a pipe lining, it is used only to a limited extent as unsupported pipe.

Extruded rigid PVC pipe is now produced at an annual rate of 1,100,000 tons and accounts for over 80% of all thermoplastic pipe produced. Although most of this and PVDC pipe is used as water pipe, considerable amounts of PVC pipe are used for transporting corrosive liquids.

PVC and other rigid pipe may be threaded and joined by threaded fittings. However, inasmuch as the threading process reduces the thickness of the pipe, some corrosion engineers prefer to use pipe that is flared to form a bell at one end so that the end of another length of pipe can serve as the spigot. Short lengths of larger-sized pipe are also used to join two lengths of pipe.

The bell and spigot or sleeves of PVC and other solvent-softened pipe are adhered by use of a solution of the plastic in a good solvent. Ketone solvents are used for solvent-welding PVC pipe (SWP). Providing the surface is clean and the temperature is above 25°C, the solvent will penetrate below the pipe surface and provide a strong joint.[12,13]

The chemical and temperature resistance of rigid PVC pipe may be reduced slightly when flexibilizing agents are present. However, properly supported PVC pipe with provision for expansion is resistant to aqueous solutions of inorganic acid, alkalies, and salts at temperatures up to 50°C.[14]

Although cellulose acetate butyrate (CAB) pipe is not as resistant to corrosives as PVC, SWP CAB has been used satisfactorily for conveying natural gas and potable water. CAB pipe has also been used for conveying sour crude petroleum.

Over 165,000 tons of acrylonitrile–butadiene–styrene terpolymer (ABS) pipe and threaded fittings are used annually. Most of this threaded pipe is used as drain waste and vent (DWV) pipe in residential construction. ABS pipe is more resistant to impact than PVC and more resistant to corrosives than CAB, but slightly less resistant to corrosives than rigid PVC pipe.

Polymethyl methacrylate (PMMA) pipe is available as unpigmented pipe. It has been used to a limited extent when transparency is essential.

Inasmuch as LDPE pipe is available in long coils, it requires few joints, and these are usually mechanical. Fifteen thousand tons of LDPE pipe and conduit are used annually in the United States. LDPE pipe has been used for conveying potable water in small water systems and as an insert to permit the use of corroded metal pipe. LDPE is resistant to aqueous solutions of nonoxidizing acids, salts, and alkalies, as well as polar solvents, at temperatures up to 35°C. The physical and chemical properties of linear low density polyethylene pipe (LLDPE) are similar to those of LDPE. Pipe that is useful at higher temperatures has been produced by extruding a mixture of polyethylene and an alkoxysilane. The extruded pipe is crosslinked by exposure to live steam.[14]

Over 240,000 tons of HDPE pipe is used annually. Over 60% of this pipe is corrugated and thus is available as coils in long lengths.[15] HDPE pipe is more rigid than LDPE pipe and therefore may be used at slightly higher temperatures. The chemical resistance of HDPE is slightly better than that of LDPE pipe.

Large diameter HDPE pipe is being used as an insert for the repair of old metal pipe[16] and for conveying oil tanker ballast from an offshore plant in Alaska. HDPE pipe wrapped with steel armor and jute is being used to transport effluent from sewer outfalls to offshore.[17]

About 10,000 tons of polypropylene (PP) pipe is used annually in the United States. The chemical resistant properties of this rigid pipe are similar to those of HDPE.

Sections of PP pipe are being thermally welded together by passing an electric current through a copper coil inserted in the space between the bell and spigot of the pipe.[18] Extruded polybutene pipe may be used for hot aqueous corrosive solutions.

Polyfluorocarbon pipe has excellent resistance to hot solutions of acids, alkalies, and salts, as well as solvents, but is more costly than PVC or ABS pipe. Polyvinylidene fluoride (PVDF) pipe has been used successfully for heat exchangers and for conveying corrosive liquids.[19]

Fiberglass-filled polyacetal has been used as pipe, fittings, and valves.[20] Other engineering plastics such as nylons, polyethylene terephthalate (PET), polybutylene terephthalate (PBT), polycarbonate, polyphenyl sulfones, polyphenylene oxide (PPO), silicone, and polyphenylene sulfide (PPS), have also been used to a limited extent as extruded pipe. These pipes have excellent resistance to corrosives and may be used at relatively high temperatures (100°C). Comprehensive reports on the installed cost of plastic pipes[21] and centrifugal pumps[22] are available. Molded thermoplastic valves are shown in Figures 11.1 and 11.2.

Figure 11.1. Thermoplastic valve. (Courtesy of Celanese Piping Systems)

Figure 11.2. Thermoplastic valve. (Courtesy of Celanese Piping Systems)

THERMOPLASTIC STRUCTURES

Chemical processing equipment such as ducts was fabricated from sheets of polyvinylidene chloride (Saran) in the United States in the 1930's. These thermoforming and thermowelding techniques were adapted to the fabrication of rigid PVC structures in Germany during World War II and reintroduced in the United States by the technical team that had visited post-war Germany.[23]

Thick PVC sheet made by the hot press lamination of calendared PVC thin sheets was fabricated into ducts, hoods, etc., in the United States in the early 1950's.[24] These fabrication techniques used for rigid PVC have been extended to other thermoplastic sheet such as LDPE, HDPE, PP, ABS, PMMA, and nylon. Extruded PVC and other thermoplastic sheet has replaced that produced by the hot lamination process.

PVC, polyolefin, and nylon sheets are readily thermoformed into the desired shapes and welded by inserting a hot round rod of the plastic between the bevelled edges of two sheets. The welding process is repeated several times to build up a strong oversized welded joint.[24,25]

It is customary to use 0.64 cm thick sheets for large ductwork but 0.32 cm thick sheets may be used for smaller ductwork. A 1 cm wide flange of thicker sheet is welded on the ends of each section. A plasticized PVC or elastomeric gasket is placed between the flanges and the latter are connected using alloy metal bolts.

ABS and PMMA sheets are also thermoformed but the shaped sheets are solvent-welded using a solution of the plastic in a good solvent. This process is similar to that described earlier in this chapter for SWP pipe.[26] The fabrication of complex structures has been simplified because of the availability of large plastic pipe, large extruded plastic sheet, and improvements in thermoforming techniques.

Thermoforming dates back to ancient times when Egyptians softened tortoise shells in hot water and made various shaped objects by forming and cooling the shells. A small amount of celluloid sheet was thermoformed in the nineteenth century but most of the early progress resulted from the thermoforming of acrylic sheet for aircraft canopies in the 1930's and 1940's.[27]

Over 124,000 tons of PVC film and sheet is extruded annually in the United States but only about 15,000 tons of this sheet are thermoformed. Much smaller quantities of the sheet are used for fabricating ducts, self supporting vessels, and other corrosion resistant structures.

Thermoplastic structures are fabricated from both unmodified PVC (Type I) and high impact modified PVC (Type II). The former may be used at temperatures as high as 65°C, whereas the latter should not be used at temperatures above 60°C. Both have excellent resistance to corrosive fumes. The impact resistance of Type II is 65 cm kg/cm of notch, whereas that of Type I PVC is 5 cm kg/cm.

Over 50,000 tons of HDPE and PP sheet are extruded annually. Only a small amount of these sheets is used for the fabrication of chemical resistant structures. Both HDPE and PP may be thermoformed and thermal-welded. These polyolefins have excellent resistance to fumes from oxidizing and nonoxidizing acids, alkalies, and salts. Some corrosion problems have been solved by the use of molded HDPE pails and drums. Molded PP is also widely used for battery cases.

Acrylic sheet (PMMA) is produced both by continuous casting and extrusion at an annual rate of 100,000 tons. Most of this sheet is used as windshields, canopies, windows, and signs.

Cellulose acetate film and sheet are widely used as membranes for reverse osmosis, ultrafiltration, and electrodialysis.[28] A perfluorosulfonic acid copolymer film reinforced by polytetrafluoroethylene is used as a diaphragm in chlorine/caustic electrolytic cells.

Over 70,000 tons of ABS sheet is produced annually in the United States. However, most of the sheet is thermoformed to produce luggage and automotive parts. Sheet and film based on fluorocarbons, nylons, polycarbonate, thermoplastic polyester, polyimides, polystyrene, polyphenylene sulfone, and polyurethane are also available but have been used only to a limited extent for the fabrication of plastic structures.

Information on plastic films and sheet has been tabulated.[29]

REFERENCES

1. R. B. Seymour, in *Short Course on Corrosive Problems in the Process Industries,* I. Cornet, Ed., University of California at San Francisco, San Francisco, Nov., 1955.
2. R. B. Seymour and R. H. Steiner, Chapter 8 in *Plastics for Corrosion Resistant Applications,* Reinhold Publishing Co., New York, 1955.
3. R. B. Seymour, *Engineering Materials Handbook,* C. L. Mantell, Ed., McGraw-Hill, New York, 1958.
4. R. B. Seymour, Chapter 15 in *Symposium on Corrosion Fundamentals,* Ade S. Brasunas and E. E. Stansburg, Eds., University of Tennessee Press, Knoxville, TN, 1956.
5. J. C. Martin, in *Managing Corrosion with Plastics,* National Association of Corrosion Engineers, Houston, TX, 1977.
6. R. M. Webster, in *Managing Corrosion with Plastics,* National Association of Corrosion Engineers, Houston, TX, 1977.
7. E. J. Wolf and R. Heim, in *Managing Corrosion with Plastics,* National Association of Corrosion Engineers, Houston, TX, 1977.
8. T. R. Morena and W. Freundlich, in *Managing Corrosion with Plastics,* National Association of Corrosion Engineers, Houston, TX, 1977.
9. J. M. Ayres, in *Managing Corrosion with Plastics,* National Association of Corrosion Engineers, Houston, TX, 1977.
10. R. B. Seymour, *Corrosion* **11**(4), 70 (1955).
11. R. B. Seymour, *Chem. Eng.* **62**(6), 280 (1955).
12. R. B. Seymour and R. H. Steiner, Chapter 17 in *Plastics for Corrosion Resistant Applications,* Reinhold Publishing Co., New York, 1955.
13. *Handbook of PVC Pipe,* Uni-Bell Plastic Pipe Association, Dallas, TX, 1978.
14. E. C. Szamborski and R. A. Marcelli, *SPE Technical Papers,* Vol. XXII, 34th Ann. Tech. Conf., Atlantic City, NJ, April 1976, pp. 324–326.
15. B. Williamson, *Plastics and Rubber* (785), 21 (1979).
16. B. Williamson, *Plastics and Rubber* (766), 1 (1977).
17. D. R. Wells, *ASCE* **46**(9), 62 (1976).
18. I. Miller, *Aus. Plastics and Rubber* **30**(3), 4 (1979).
19. R. K. Johannes, *Plast. Eng.* **34**(7), 24 (1978).
20. N. C. Baldwin, *Plast. Eng.* **34**(5), 9 (1978).
21. J. Yamartino, *Chem. Eng.* **85**(26), 138 (1978).
22. E. Margus, *Chem. Eng.* **85**(5), 215 (1977).
23. R. B. Seymour and R. H. Steiner, Chapters 15 and 16 in *Plastics for Corrosion Resistant Applications,* Reinhold Publishing Co., New York, 1955.
24. J. L. Huscher, *Materials and Methods* **39**(6), 119 (1954).
25. G. S. Laaff, *Chem. Eng. Progr.* **50**(6), 275 (1954).
26. R. B. Seymour, in *Plastics in Building,* Building Research Institute, Washington, DC, 1955.

27. W. K. McConnell, Chapter 12 in *SPI Plastics Engineering Handbook,* J. Frados, Ed., Van Nostrand Reinhold Co., New York, 1976.

28. G. S. Kirshenbaum, *Polymer News* **3,** 324 (1977).

29. S. Gross, *Modern Plastics* **55**(104), 640 (1979).

CHAPTER 12

Reinforced Plastic Pipe, Tanks, and Structures

Whereas thermoplastic pipe and other structures were developed specifically for chemical resistant applications, reinforced plastics were developed originally for aircraft and marine applications. Neither the fiberglass reinforcements nor the polyester resin used originally were particularly resistant to corrosives, but they have been replaced by more satisfactory materials.

The early emphasis in the 1940's was on reinforced thermosets, particularly unsaturated polyesters, and reinforced thermoplastics were not introduced commercially until the mid-1950's. Over 110,000 tons of reinforced thermoset plastics are used annually for the fabrication of chemical resistant equipment.

REINFORCED THERMOPLASTIC PIPE

Most thermoplastics are available as filled compounds that may be extruded to produce filled thermoplastic pipe. Many of the pipes with fillers having high aspect ratios (l/d) are stronger and more heat resistant than the unfilled pipe. Providing the filler itself is chemically resistant, the chemical resistance of the pipe will be superior to that of the unfilled pipe. Best results are obtained when coupling agents such as silanes or organotitanates are present to assure a good interfacial bond.

Superior physical properties are obtained when the thermoplastic pipe is reinforced with about 30% glass fibers. Even better properties are obtained with graphite filaments.

The most heat resistant filled thermoplastic pipe is obtained by extruding mixtures of reinforcing fibers and engineering plastics. Among the latter are nylons, polyacetals, thermoplastic polyesters, polycarbonate, polyphenylene sulfones, polyphenylene oxide, polyphenylene sulfide, polyfluorocarbons, and thermoplastic polyurethanes.

Although over 115,000 tons of reinforced thermoplastics are produced annually in the United States, only a small amount is extruded as pipe. The principal filled thermoplastics used are as follows: polypropylene (33%), thermoplastic polyesters (21%), nylon (19%), styrene polymers (15%), and polycarbonate (4%). The other reinforced engineering plastics account for less than 10% of the total.

Additional information on reinforced thermoplastics may be found in the references listed at the end of this chapter.[1-7]

REINFORCED THERMOSET PIPE

One of the simplest reinforced pipe is reinforced plastic mortar pipe (RPMP) consisting of ground sand-filled isophthalic polyester resin reinforced by filament-wound fiberglass and used primarily as sewer pipe, including that inserted into deteriorating sewer lines.[8] Some low-pressure-reinforced plastic pipe has been made by hand lay-up techniques over a thermoplastic pipe such as PVC (skinning) or over a removable mandrel. Better pipe has been made by centrifugal casting and superior pipe has been produced by filament-winding and pultrusion.

In the centrifugal casting technique, a mixture of chopped fibers and catalyzed prepolymer is heated and rotated within a mandrel. In the filament-winding process, a filament is passed through a bath of resin and catalyst, wound on a mandrel, and cured by heating. In the pultrusion process a bundle of filaments is passed through a bath of resin and catalyst and then through a hot circular die to produce a pipe on a continuous basis.[9] It is customary to use chemical resistant glass fiber such as E or S glass for making chemically resistant pipe.

Fittings and sleeve joints must be molded separately.[10] These fittings may be threaded or cemented on the plastic pipe by use of a polyester premix cement using butt or strap joints.[11] The cement may also be used for making press-molded flanges for reinforced plastic pipe systems.[12] Reinforced plastic tank and ductwork in plants are shown in Figures 12.1 and 12.2.

General-purpose unsaturated polyesters are used for making boats and other structures, but more chemically resistant resins are preferred for pipe used to convey corrosive liquids. These chemically resistant resins include: orthophthalic and isophthalic polyesters, bisphenol-A-fumarate (Atlac 382), epocryl methacrylic esters (vinyl esters, Epocryl, Derakane), and a new vinyl ester resin.[13,14] These polyesters are resistant to acid and acid salts but

Figure 12.1. Reinforced plastic tank and ductwork in a petrochemical plant. (Courtesy of A. O. Smith-Inland Inc.)

Figure 12.2. Reinforced polyester tank and ductwork in a chemical plant. (Courtesy of Ceilcote Co.)

are not satisfactory for service with strong alkaline solutions. A large reinforced polyester pipeline, exhaust system, fume duct, and tank are shown in Figures 12.3–12.6.

A study of long term service of filament-wound plastic pipe has shown that a vinyl ester pipe is satisfactory for conveying 80 percent sulfuric acid at 80°C but is not satisfactory for use with 98 percent sulfuric acid (an oxidizing acid) at room temperature.[15] Fiberglass-reinforced epoxy resin pipe made by centrifugal casting or by the filament-winding process is resistant both to nonoxidizing acids and alkalies, up to 80°C.[16]

When the pipe wall is sufficiently thick, filament-wound fiberglass-reinforced furan pipe (RTFP) is resistant to nonoxidizing acids, alkalies, salts, and organic solvents. Furan pipe is not satisfactory for conveying sodium hypochlorite solutions.[17] Reinforced plastic pipe carrying strong corrosives at temperatures above 80°C should be inspected every 18 months.[18] In some instances, defects may be repaired by use of thermosetting plastic ce-

Figure 12.3. Large reinforced polyester pipeline. (Courtesy of Amoco Chemical Corp.)

Figure 12.4. Reinforced polyester exhaust system. (Courtesy of Amoco Chemical Corp.)

ments.[19] Standards have been established for filament-wound (ASTM D-2996), centrifugally cast (ASTM D-2997), sewer pipe (ASTM D-3262) and other aspects of RTRP pipe.

REINFORCED PLASTIC TANKS

Small reinforced plastic vessels can be molded from sheet molding compounds (SMC) and bulk molding compounds (BMC) which contain short fibers, fillers, resin, and catalyst. However, most vessels are fabricated by spray-up techniques as well as those cited for reinforced pipe using polyester, vinyl ester, epoxy, and furan resins.[20] Sections of tanks may also be fabricated and bolted together as are flanged pipe sections.

It is important to design tanks with rounded corners (knuckles) whenever possible and to build up FRP rings or brackets all around the tank.[21] The controlling property for mechanical design of FRP vessels is strain which is a sound basis for establishing design stress.[22]

Standards have been established for self-supporting structures (ASTM C-582) and filament-wound-reinforced tanks (ASTM D-3299) by Subcom-

Figure 12.5 Reinforced polyester fume duct. (Courtesy of Amoco Chemical Corp.)

Figure 12.6. Reinforced polyester tank. (Courtesy of Amoco Chemical Corp.)

mittee XVII of ASTM Committee D-20 and many more are under development.[23] Strain gauge data have been used to assure correct design of large FRP pressure vessels.[24]

A structural resin should never be subjected to temperatures higher than 10°C below its heat deflection temperature. Data on flexibility factor, flexural modulus, and tensile strength of the structural resin can be used to assure good strength and flexibility of the composite.[25] Stress intensification factors (sif's) and flexibility factors may be obtained empirically from FRP bends and tees, and used for the design of vessels in chlorine manufacturing plants.[26]

As a result of immersion tests on cutout coupons with exposed edges in solvents, it was concluded that an epoxy novolac vinyl ester resin was superior to conventional C R polyester resins.[27] The chemical resistance of improved furan resins (Qua Corr 1001) is comparable to that of corrosion resistant polyester and vinyl ester resins.[28]

REINFORCED PLASTIC STRUCTURES

Although the use of GRP structures such as ducts and stacks may not be as critical as that of pipe and vessels, proper selection of the resin, design of the structure, and careful fabrication are important for good serviceability. There is a correlation between poor quality and poor chemical resistance and vice versa.[29]

These structures are usually fabricated by hand lay-up or spray-up of fiberglass-reinforced unsaturated polyester resins but epoxy, phenolic, furan, and vinyl ester resins are also used.[30] The latter are susceptible to stress corrosion at elevated temperature service if immersed in hydrochloric acid or strong alkalies but are acceptable for making hoods and stacks.[31] Superior resistance to corrosives has been shown for stacks and ducts fabricated from polyester resins based on trimethylpentane diol (TMPD)[32] and chlorendic anhydride.[33] The latter are more resistant to flame than general-purpose (G.P.) polyesters. Reinforced polyester stack and ductwork are shown in Figures 12.7 and 12.8.

In spite of some failures by fire, wind, and corrosives, numerous properly designed fume stacks are still in service after many years of use.[34] Specific design data for a filament-wound 37 m tall × 4 m diameter stack operating at 149°C are available.[35] Reinforced polyester stacks, tanks, waste lines, and scrubbers are shown in Figures 12.9–12.18.

Although a fiberglass-reinforced vinyl ester was selected for ductwork, propoxylated bisphenol-A-fumarate, modified bisphenol, and chlorendic anhydride polyesters were considered satisfactory for exhausting wet chlo-

Figure 12.7. A reinforced polyester stack. (Courtesy of Ashland Chemical Corp.)

rine at 85°C.[36] Reinforced bisphenol-A and isophthalic polyester are considered adequate for scrubbers in the aluminum industry.[37]

The GRP housing of a centrifugal fan is less complicated than the fan itself, but it must be properly designed to avoid mechanical failure and fire damage.[38,39] The use of GRP equipment in the paper industry has resulted in the production of paper with higher brightness levels.[40]

Additional information on reinforced plastics may be found in the references at the end of this chapter.[41–45] New information is also reported each year at the Annual Meeting of the SPI Reinforced Plastics/Composites Institute.

Figure 12.8. Reinforced polyester ductwork. (Courtesy of Ashland Chemical Corp.)

Figure 12.9. Reinforced polyester fume stack. (Courtesy of Amoco Chemical Corp.)

Figure 12.10. Reinforced polyester fume stack. (Courtesy of Amoco Chemical Corp.)

Figure 12.11. Reinforced polyester tank. (Courtesy of Amoco Chemical Corp.)

Figure 12.12. Reinforced polyester storage tank. (Courtesy of Amoco Chemical Corp.)

Figure 12.13. Reinforced polyester storage tanks. (Courtesy of Amoco Chemical Corp.)

Figure 12.14. Reinforced polyester treatment tank. (Courtesy of Amoco Chemical Corp.)

Figure 12.15. Large vertical reinforced polyester tank. (Courtesy of Ceilcote Co.)

Figure 12.16. Large horizontal reinforced polyester storage tank. (Courtesy of Ceilcote Co.)

Figure 12.17. Reinforced polyester wasteline. (Courtesy of Amoco Chemical Corp.)

Figure 12.18. Reinforced polyester scrubbers. (Courtesy of Amoco Chemical Corp.)

REFERENCES

1. P. Hamed and A. Y. Coran, Chapter 3 in *Additives for Plastics,* Vol. 2, Raymond B. Seymour, Ed., Academic Press, New York, 1978.
2. H. Katz and L. Ehenreich, Chapter 3 in *Additives for Plastics,* Vol. 2, Raymond B. Seymour, Ed., Academic Press, New York, 1978.
3. H. Katz and J. V. Milewski, *Handbook of Fillers and Reinforcements for Plastics,* Van Nostrand-Reinhold Co., New York, 1978.
4. J. E. Theberg, *Modern Plastics* **55**(104), 140 (1979).
5. W. V. Titow and B. J. Lanham, *Reinforced Thermoplastics,* John Wiley & Sons, New York, 1975.
6. M. P. Wagner, Chapter 2 in *Additives for Plastics,* Vol. 1, Raymond B. Seymour, Ed., Academic Press, New York, 1978.
7. M. P. Wagner and M. Q. Fetterman, Chapter 4 in *Additives for Plastics,* Vol. 2, Raymond B. Seymour, Ed., Academic Press, New York, 1978.
8. B. J. Schrock and G. Rangel, in *SPI Reinforced Plastics/Composites Institute Technical Review,* Society of the Plastics Industry, New York, 1976.
9. J. Frados, Chapter 9 in *Plastics Engineering Handbook,* J. F. Frados, Ed., Van Nostrand-Reinhold Co., New York, 1976.

10. A. R. Henning and G. B. Cuming, Chapter 11 in *Glass Reinforced Plastics,* P. Morgan, Ed., Iliffe and Sons, London, 1954.
11. G. A. Escher and W. B. MacDonald, in *Managing Corrosion with Plastics,* National Assn. of Corrosion Engineers, Houston, TX, 1977, p. 276.
12. D. E. Fidder and J. Lazarow, in *SPI Reinforced Plastics/Composites Institute Technical Review,* SPI, New York, 1976.
13. P. J. Lewandowski, E. C. Ford, D. M. Longnecker, A. J. Restaino, and J. P. Burns, in *SPI Reinforced Plastics/Composites Institute Technical Review,* SPI, New York, 1976.
14. F. Jacobs, L. L. Brass, Proc. 35th Ann. Conf. SPI Reinf. Plast./Comp. Inst., 1980.
15. K. J. Oswald, in *Managing Corrosion Problems with Plastics,* National Assn. of Corrosion Engineers, Houston, TX, 1977, p. 240.
16. D. W. Bordner, *SPI Reinforced Plastics/Composites Institute Technical Review,* SPI, New York, 1976.
17. E. D. Edmisten, K. D. Maguire, and R. A. Currieo, in *Managing Corrosion with Plastics,* National Assn. of Corrosion Engineers, Houston, TX, 1977. p. 302.
18. J. L. Cooney, in *Managing Corrosion with Plastics,* National Assn. of Corrosion Engineers, Houston, TX, 1977, p. 98.
19. D. R. Dean and G. T. Sikes, in *Managing Corrosion with Plastics,* National Assn. of Corrosion Engineers, Houston, TX, 1977, p. 57.
20. M. A. Rehm, *Modern Plastics* **55**(104), 142 (1979).
21. J. C. Lazarow, in *Managing Corrosion with Plastics,* National Assn. of Corrosion Engineers, Houston, TX, 1977, p. 18.
22. H. E. Atkinson, in *Managing Corrosion with Plastics,* National Assn. of Corrosion Engineers, Houston, TX, 1977, p. 223.
23. J. S. McDermott, in *Managing Corrosion with Plastics,* National Assn. of Corrosion Engineers, Houston, TX, 1977, p. 185.
24. R. Pace and G. Brown, in *Managing Corrosion with Plastics,* National Assn. of Corrosion Engineers, Houston, TX, 1977.
25. H. R. Edwards, *SPI Reinf. Plast./Comp. Inst. Tech. Rev.,* SPI, New York, 1976.
26. R. C. Roberts, in *Managing Corrosion with Plastics,* National Assn. of Corrosion Engineers, Houston, TX, 1977, p. 270.
27. W. W. McClellan, T. F. Anderson, and R. F. Stavinoka, in *SPI Reinf. Plast./Comp. Inst. Tech. Rev.,* SPI, New York, 1976.
28. W. R. Schnell, in *Managing Corrosion with Plastics,* National Assn. of Corrosion Engineers, Houston, TX, 1977.
29. R. E. McMahon and R. J. Lewandowski, in *SPI Reinf. Plast./Comp. Inst. Tech. Rev.,* SPI, New York, 1976.
30. P. J. Gegner, *Chem. Tech.* **9,** 676 (1979).
31. H. Hojo and H. Koyana, Proc. Int. Conf. Org. Coat. Sci. Technol. 3, 418 (1977).
32. S. L. Hillman, M. Palmer, and J. H. Davis, *Plast. Eng.* **35**(6), 32 (1979).
33. A. S. Wood, *Modern Plastics* **56**(2), 44 (1979).
34. A. K. Shadduck, in *Managing Corrosion with Plastics,* National Assn. of Corrosion Engineers, Houston, TX, 1977, p. 252.
35. D. L. Killam, in *Managing Corrosion with Plastics,* National Assn. of Corrosion Engineers, Houston, TX, 1977, p. 145.

36. D. E. Larkin, in *Managing Corrosion with Plastics,* National Assn. of Corrosion Engineers, Houston, TX, 1977, p. 143.

37. B. Hebnes, in *Managing Corrosion with Plastics,* National Assn. of Corrosion Engineers, Houston, TX, 1977, p. 138.

38. C. J. Trichler, in *Managing Corrosion with Plastics,* National Assn. of Corrosion Engineers, Houston, TX, 1977, p. 279.

39. T. Kubik, in *Managing Corrosion with Plastics,* National Assn. of Corrosion Engineers, Houston, TX, 1977, p. 24.

40. A. F. Greenfield, in *Managing Corrosion with Plastics,* National Assn. of Corrosion Engineers, Houston, TX, 1977, p. 86.

41. H. Bernstein, Symposium on Technology of Reinforced Plastics, M.I.T., Cambridge, MA, June, 1961.

42. P. Morgan, *Glass Reinforced Plastics,* Iliffe and Sons, London, 1958.

43. S. S. Oleesky and J. G. Mohr, *SPI Handbook of Reinforced Plastics,* Van Nostrand-Reinhold, New York, 1964.

44. B. Parkyn, Ed., CRC Press, Cleveland, OH, 1970.

45. J. Seavuzzo, Seminar on Reinforced Plastics, University of Wisconsin, Madison, WI, May, 1962.

CHAPTER 13

Selection Guide: Typical Thermal and Physical Properties of Commercial Plastics

If a plastic material is to be used at temperatures above room temperature (23°C), one must know its thermal properties. The heat deflection temperature under a flexural load of 264 psi as measured by ASTM test D-668 is a useful indication of the tendency of a plastic to sag under load. However, the safe temperature for continuous heat is usually a few degrees lower than the heat deflection temperature.

Since the expansion of most plastics when heated is greater than that of metals, it is important to know the coefficient of linear thermal expansion. Provisions must be made to allow for this expansion if the temperature of the plastic material is increased under normal operations.

If the plastic component is under a compressive load in service, the compressive strength is important. The flexural strength is also important if the plastic component is subjected to flexural deformation (bending).

Impact strength is a measure of toughness. Notches and sharp corners should be avoided in the design of plastic components. However, the Izod impact test, which is the most widely used impact test, is a measure of resistance to impact of a notched specimen. In general, a plastic material having an impact strength of 4 ft lb/in. (213.5 cm·*N*/cm) of notch is tougher than one with a value of 2 ft lb/in. (107 cm·*N*/cm) of notch but it is not twice as tough!

The tensile strength and elongation under tension are widely reported properties. These values are important if the plastic component is under tension in service. However, in most instances, compressive and flexural strengths and moduli are more important. The ratio of tensile strength to elongation or modulus is a measure of stiffness.

Indentation or scratch hardness may be important in certain applications of plastics. The specific gravity value is important economically and in the determination of the weight of a plastic component.

Typical values for the ten properties discussed above are listed for specific commercial plastics shown in alphabetical order. A separate table of characteristics is included (after the conclusion section of this chapter) for each, as well as a brief description of production process and uses.

ABS (ACRYLONITRILE–BUTADIENE–STYRENE TERPOLYMER)

ABS thermoplastics are usually made by the copolymerization of acrylonitrile, butadiene, and styrene. The toughness and other physical properties of these commercial products will vary with composition. ABS has fair resistance to weathering and is used as thermoformed sheet and/or extruded pipe and profiles (Table 13.1).

ACETAL

Acetal plastics are available as capped homopolymers (Delrin) and copolymers of formaldehyde (Celcon). These clear engineering thermoplastics are characterized by high modulus and good impact strength over a wide temperature range (−40 to 110°C). Pump impellers and valve parts are injection molded from acetals (Table 13.2).

ACRYLICS

Acrylic plastics, namely, polymers of alkyl methacrylates such as methyl methacrylate, and polymers of alkyl acrylates such as ethyl acrylate, or copolymers of these esters, are clear products with excellent transparency. These thermoplastics, which are available as cast or extruded sheet, are used as thermoformed articles and pipes (Table 13.3).

ALKYDS

Alkyds are thermosets that are cured by free radical polymerization. They are available as molded products and as protective coatings (Table 13.4).

ALLYLIC PLASTICS

Allyl diglycol carbonate (CR 36) is available as a cast sheet. The most widely used allylic plastics are products produced by molding diallyl phthalate polymers (DAP). These are usually available as silica or glass-filled plastics (Table 13.5).

AMINO PLASTICS

The polymers produced by the condensation of formaldehyde with urea (UF) and melamine (MF) are called amino plastics. Both may be used as baked coatings or as cellulose-filled molding resins (Table 13.6).

ARAMIDS

Aramids are aromatic polyamides characterized by excellent resistance to heat and high strength (Table 13.7).

CELLULOSICS

Cellulose is a renewable resource and one of the classical reactants for the production of plastics. However, whereas wood continues to be used to a limited extent in corrosive environments, the use of cellulosic ethers such as ethylcellulose and cellulosic esters such as cellulose acetate is limited. These thermoplastics have limited use as coatings, sheet, extruded pipe, and injection molded articles (Table 13.8).

EPOXY RESINS (EP)

Classical epoxy resins produced by the condensation of bisphenol-A and epichlorohydrin are cured at room temperature by polyamines and at elevated temperatures by cyclic anhydrides. These thermoset resins are used as protective coatings, castings, and molded plastics. All of these compositions may be reinforced with reinforcing agents such as fiberglass (Table 13.9).

FLUOROPLASTICS

Fluoroplastics are heat resistant thermoplastics consisting of polytetrafluoroethylene (PTFE, Teflon), polychlorotrifluoroethylene (PCTFE, Kel F), polyvinylidine fluoride (PVDF, Kynar), and polyvinyl fluoride (PVF, Tedlar). More flexible and more readily fabricated copolymers are also available. These are: FEP, a copolymer of tetrafluoroethylene and hexafluoropropylene; PE-CTFE, a copolymer of chlorotrifluoroethylene and ethylene; and PETFE, a copolymer of tetrafluoroethylene and ethylene. Although polytetrafluoroethylene is insoluble in solvents and is difficult to mold, the lower melting polymers and copolymers are more soluble and more readily molded and extruded. Fluoroplastics are used as coatings, extruded pipe, and molded articles (Table 13.10).

FURAN PLASTICS

Furan plastics are dark-colored products based on the polymerization products of furfuryl alcohol or furfural. The liquid resins are usually mixed with a carbonaceous or siliceous filler and catalyst, polymerized *in situ*, and used as corrosive resistant mortars (Table 13.11).

NYLONS

The original nylon produced by Carothers in the 1930's was a polyamide obtained by the condensation of adipic acid [$HOOC(CH_2)_4COOH$) and hexamethylenediamine [$H_2N(CH_2)_6NH_2$] and designated as nylon-66 to indicate the number of carbon atoms in each of the reactants. Nylons produced from higher molecular weight dicarboxylic acids are more flexible and more resistant to moisture. Polyamides such as nylon-6 are produced by the self-condensation of alpha-amino-omega-carboxylic acids such as that obtained from caprolactam.

All nylons are classified as engineering plastics. They may be molded as unfilled resins, fiberglass-reinforced resins, or as resins filled with polytetrafluoroethylene or molybdenum disulfide. The former are used as gunstocks, pipe fittings, or sheet and the latter are used as gears, cams, and bearings (Table 13.12).

PHENOLICS

Phenolic plastics were produced by Baekeland in the early 1900's by the condensation of phenol and formaldehyde and are still produced today. Novolak resins are produced by condensation under acidic conditions and resole resins are produced by condensation under alkaline conditions. Most molded phenolic plastics are filled novolak resins. Resole resins are used for castings, mortars, and laminates (Table 13.13).

POLY(AMIDE-IMIDE)

Poly(amide-imide) is a high performance thermoplastic produced by the condensation of trimellitic dianhydride and various diamines. This injection-molded plastic is used on faucets, pumps, valves, and turbines (Table 13.14).

POLYARYL ESTERS

There are several commercial thermoplastic polyaryl esters including the polyarylates, polyethylene terephthalate (PET), polybutylene terephthalate (PBT), and copolymers of PET and PBT. Molded parts from polyaryl esters produced by the condensation of phthalic acid and bisphenol-A have been used as electrical parts and solar energy collectors. PET has been used for many years as a fiber and film and is now used for the blow-molding of bottles and other containers. PBT, which is more readily molded than PET, is used for molded pump housings, impellers, valves, gears, and nozzles (Table 13.15).

POLYCARBONATES

Polycarbonates may be produced by the condensation of bisphenol-A and phosgene. These clear tough thermoplastics are used for blow-molded containers and extruded sheet (Table 13.16).

POLYESTERS (UNSATURATED)

The unsaturated polyester plastics usually consist of fiberglass-reinforced polyesters produced by the *in situ* copolymerization of a solution of an unsaturated polyester in styrene. Partially polymerized sheet composites (SMC) and doughs (BMC) may be compression molded. Catalyzed-resin-impregnated glass fiber or polyester cements harden (polymerize) at ordinary temperatures. These high volume plastics are used as tanks, sinks, bathtubs, trays, pump housings, flooring, and laminated products (Table 13.17).

POLYETHYLENES

Polyethylene is available as high density (HDPE), low density (LDPE), and linear low density (LLDPE) polyethylene. The American Society for Testing Materials (ASTM) has divided these polymers and their blends into the following types according to their specific gravity:

Type I	0.910–0.925
Type II	0.926–0.940
Type III	0.941–0.965

Figure 13.1. Polyethylene photographic developing tanks. (Courtesy of Nalge Co.)

Figure 13.2. Polyethylene rectangular tank. (Courtesy of Nalge Co.)

LDPE is characterized by stress cracking in the presence of solvents. HDPE is stiffer and more heat resistant than LDPE. High-molecular-weight polyethylene (UHMWPE) is characterized by outstanding strength and toughness (Table 13.18). Cylindrical and rectangular polyethylene tanks are shown in Figures 13.1 and 13.2.

POLYIMIDES

Both thermoplastics and thermoset polyimides are commercially available. The aromatic thermoplastic polyimides may be compression-molded to produce heat resistant articles such as piston rings, valve seats, and thrust washers. The more heat resistant thermoset polyimides may be compression-molded to produce jet and turbo engine parts. These polyimides are also available as films and coatings (Table 13.19).

POLYMETHYLPENTENE

Polymethylpentene (TPX) is one of the lightest of the solid plastics. It is also characterized by good transparency and fair resistance to heat. This polymer may be injection-molded to produce laboratory utensils, lenses, containers, and electrical equipment (Table 13.20).

POLYPHENYLENE OXIDE

Polyphenylene oxide (PPO) is produced by the copper-ion-oxidative coupling of cresols. PPO blends with polystyrene are used as film processing tanks, electrical applications, pump parts, and machine housings (Table 13.21).

POLYPHENYLENE SULFIDE

Polyphenylene sulfide (PPS) is a high-melting heat-stable solvent-resistant aromatic polymer. Injection and compression-molded filled and unfilled PPS have been used for electrical applications, pumps, under-the-hood applications, and cooking appliances (Table 13.22).

POLYPROPYLENE

Polypropylene (PP) is a translucent hydrocarbon plastic with a low specific gravity (0.90) and fair resistance to heat. Injection-molded PP is used in appliances, electrical applications, ducts, and housewares. PP filaments are used for rope, netting, and carpets. Atactic PP is a viscous liquid but isotactic PP, the commercial product, is a hard plastic (Table 13.23).

POLYSTYRENE

Polystyrene (PS) is a colorless brittle solid with excellent electrical properties but poor resistance to heat. High impact PS may be an elastomer-modified PS or a copolymer, and it has good resistance to impact. PS is used for furniture, housewares, packaging, film, and electrical applications (Table 13.24).

POLYSULFONES

The three commercially available sulfone polymers are polysulfone, polyphenyl sulfone, and polyether sulfone. These heat resistant aromatic polymers have excellent electrical and thermal properties and are used as circuit boards, micro ovenware, valves, motor components, and for electrical applications (Table 13.25).

POLYURETHANES

Although polyurethanes (PU) are used primarily as flexible and rigid foams, they are also available as elastomers, castings, adhesives, and coatings. These products are characterized by excellent wear resistance and toughness and are used as wheels, pulleys, gears, pump impellers, conveyor belts, and flooring (Table 13.26).

POLYVINYL CHLORIDE

Vinyl chloride polymers (PVC) are available as rigid PVC, plasticized PVC, chlorinated PVC (PVDC), and copolymers such as those with vinyl

Figure 13.3. PVC pipe and valve system. (Courtesy of B. F. Goodrich Chemical Group)

acetate. PVC may be injection-molded or extruded and a suspension of PVC in a liquid plasticizer such as dioctyl phthalate may be used as a plastisol that becomes solid when heated at 160°. Rigid PVC is used as pipe, tanks, and blow-molded containers. Plasticized PVC is used as flexible tubing, gaskets, wire insulation, film, and sheeting (Table 13.27). A rigid PVC pipe and valve system is shown in Figure 13.3.

SILICONES

In contrast to organic polymers, which usually consist of chains of covalently bonded carbon atoms, silicones are inorganic polymers with a siloxane (–Si–O–) backbone. Silicone plastics are available as castings (RTV), reinforced compression-molded parts, and elastomers. Silicones are characterized by excellent resistance to heat and moisture and are used as sealants, prosthetic devices, electrical components, and coatings (Table 13.28).

THERMOPLASTIC ELASTOMERS

Thermoplastic elastomers (TPE) are unvulcanized flexible molded products that contain rigid and flexible domains in the polymeric chain. TPE includes styrene–butadiene block copolymers, polyolefin blends, polyester block copolymers, and polyurethane block copolymers. These products differ in chemical structure but have similar morphology. They are used as automotive components, mechanical goods, hose, shoe soles, cable insulation, sheeting, sealants, coatings, and adhesives (Table 13.29).

CONCLUSION

The values given in the following tables are typical values that may be used as guides for selection of plastics to meet specific physical and thermal requirements. Inasmuch as the physical and thermal values will vary with compounding techniques used in processing, the user should attempt to secure useful data from the supplier of the end product.

Additional information may be found in suppliers' literature and in the suggested reading at the end of this chapter.

Table 13.1. ABS

Thermal and Physical Properties of Plastics (Typical Values)	Extrusion Grade	20% Glass-Reinforced
Heat deflection temperature @ 1820 kPa (°C)	90	100
Maximum resistance to continuous heat (°C)	80	90
Coefficient of linear expansion cm/cm/°C $\times 10^{-5}$	9.5	2.0
Compressive strength (kPa)	48,265	96,530
Flexural strength (kPa)	62,055	103,425
Impact strength (Izod: cm · N/cm of notch)	320.3	53.4
Tensile strength (kPa)	34,475	75,845
Elongation %	60	5
Hardness: Rockwell	R60	M85
Specific gravity	1.03	1.2

Table 13.2. Acetals

Thermal and Physical Properties of Plastics (Typical Values)	Homopolymer	Copolymer	25% Glass Reinforced Copolymer
Heat deflection temperature @ 1820 kPa (°C)	125	110	160
Maximum resistance to continuous heat (°C)	100	100	125
Coefficient of linear expansion cm/cm/°C $\times 10^{-5}$	10.0	8.5	5.0
Compressive strength (kPa)	106,110	110,320	117,215
Flexural strength (kPa)	96,530	89,635	193,060
Impact strength (Izod: cm · N/cm of notch)	80.1	69.4	96.1
Tensile strength (kPa)	68,950	62,055	128,600
Elongation %	30	50	3
Hardness: Rockwell	M94	M78	M79
Specific gravity	1.412	1.41	1.61

Table 13.3. Acrylics

Thermal and Physical Properties of Plastics (Typical Values)	Cast Acrylics	Acrylic-PVC Alloy
Heat deflection temperature @ 1820 kPa (°C)	95	70
Maximum resistance to continuous heat (°C)	75	60
Coefficient of linear expansion cm/cm/°C $\times 10^{-5}$	7.0	6.0
Compressive strength (kPa)	103,425	57,920
Flexural strength (kPa)	96,530	72,397
Impact strength (Izod: cm · N/cm of notch)	21.4	8.0
Tensile strength (kPa)	65,500	44,820
Elongation %	4	100
Hardness: Rockwell	M80	R100
Specific gravity	1.18	1.25

Table 13.4. Alkyds

Thermal and Physical Properties of Plastics (Typical Values)	Glass-Filled Alkyds
Heat deflection temperature @ 1820 kPa (°C)	200
Maximum resistance to continuous heat (°C)	200
Coefficient of linear expansion cm/cm/°C $\times 10^{-5}$	2.0
Compressive strength (kPa)	137,900
Flexural strength (kPa)	103,425
Impact strength (Izod: cm · N/cm of notch)	106.7
Tensile strength (kPa)	41,370
Elongation %	2
Hardness: Rockwell	80E
Specific gravity	2.1

Table 13.5. Allylic plastics

Thermal and Physical Properties of Plastics (Typical Values)	Glass-Filled Allylic Plastics (DAP)
Heat deflection temperature @ 1820 kPa (°C)	200
Maximum resistance to continuous heat (°C)	150
Coefficient of linear expansion cm/cm/°C $\times 10^{-5}$	2.0
Compressive strength (kPa)	186,000
Flexural strength (kPa)	131,000
Impact strength (Izod: cm · N/cm of notch)	106
Tensile strength (kPa)	58,000
Elongation %	4
Hardness: Rockwell	E80
Specific gravity	1.7

Table 13.6. Amino plastics

Thermal and Physical Properties of Plastics (Typical Values)	Cellulose-Filled MF	Cellulose-Filled UF
Heat deflection temperature @ 1820 kPa (°C)	150	130
Maximum resistance to continuous heat (°C)	100	75
Coefficient of linear expansion cm/cm/°C $\times 10^{-5}$	4	3
Compressive strength (kPa)	276,000	220,648
Flexural strength (kPa)	86,000	96,500
Impact strength (Izod: cm · N/cm of notch)	16	16
Tensile strength (kPa)	68,950	55,160
Elongation %	0.7	0.7
Hardness: Rockwell	M115	M110
Specific gravity	1.5	1.5

Table 13.7. Aramid

Thermal and Physical Properties of Plastics (Typical Values)	Aramid
Heat deflection temperature @ 1820 kPa (°C)	260
Maximum resistance to continuous heat (°C)	150
Coefficient of linear expansion cm/cm/°C $\times 10^{-5}$	2.6
Compressive strength (kPa)	206,850
Flexural strength (kPa)	172,325
Impact strength (Izod: cm·N/cm of notch)	74.5
Tensile strength (kPa)	120,000
Elongation %	5
Hardness: Rockwell	E90
Specific gravity	1.2

Table 13.8. Cellulosics

Thermal and Physical Properties of Plastics (Typical Values)	Ethyl Cellulose	Cellulose Acetate	Cellulose Acetate Butyrate
Heat deflection temperature @ 1820 kPa (°C)	65	65	65
Maximum resistance to continuous heat (°C)	60	60	60
Coefficient of linear expansion cm/cm/°C $\times 10^{-5}$	15.0	1.25	1.40
Compressive strength (kPa)	120,000	55,000	34,500
Flexural strength (kPa)	41,370	55,200	41,350
Impact strength (Izod: cm·N/cm of notch)	21.35	106.7	160
Tensile strength (kPa)	34,475	41,510	34,475
Elongation %	10	25	50
Hardness: Rockwell	R60	R80	R75
Specific gravity	1.1	1.3	1.2

Table 13.9. Epoxy resins

Thermal and Physical Properties of Plastics (Typical Values)	Epoxy Plastic	Glass-Filled EP	Glass-Spheres-Filled EP
Heat deflection temperature @ 1820 kPa (°C)	140	150	115
Maximum resistance to continuous heat (°C)	120	135	110
Coefficient of linear expansion cm/cm/°C $\times 10^{-5}$	2.5	2.0	2.5
Compressive strength (kPa)	120,000	206,850	82,740
Flexural strength (kPa)	124,100	103,400	41,370
Impact strength (Izod: cm · N/cm of notch)	53.4	53.4	10.6
Tensile strength (kPa)	51,710	82,740	41,370
Elongation %	5	4	1
Hardness: Rockwell	M90	M105	—
Specific gravity	1.2	1.8	0.8

Table 13.10. Fluoroplastics

Thermal and Physical Properties of Plastic (Typical Values)	PFTE	PCTFE	PVDF	PVF	PE-CTFE	PE-TFE
Heat deflection temperature @ 1820 kPa (°C)	100	100	80	90	115	120
Maximum resistance to continuous heat (°C)	250	200	150	125	100	160
Coefficient of linear expansion cm/cm/°C $\times 10^{-5}$	10	14	8.5	10	8	7
Compressive strength (kPa)	27,580	38,000		—	41,370	48,260
Flexural strength (kPa)		60,000		—	48,260	38,000
Impact strength (Izod: cm·N/cm of notch)	160	133.5		—	—	—
Tensile strength (kPa)	24,130	34,475	55,160		48,260	48,260
Elongation %	200	100	200	—	200	250
Hardness: Rockwell	D52	R80	R110	—	R95	R50
Specific gravity	2.16	2.1	1.76	1.4	1.7	1.7

Table 13.11. Furan plastics

Thermal and Physical Properties of Plastics (Typical Values)	Furan (Carbon-Filled)
Heat deflection temperature @ 1820 kPa (°C)	80
Maximum resistance to continuous heat (°C)	100
Coefficient of linear expansion cm/cm/°C $\times 10^{-5}$	7.5
Compressive strength (kPa)	69,000
Flexural strength (kPa)	34,475
Impact strength (Izod: cm·N/cm of notch)	21.4
Tensile strength (kPa)	41,370
Elongation %	1.5
Hardness: Rockwell	R110
Specific gravity	1.7

Table 13.12. Nylons

Thermal and Physical Properties of Plastics (Typical Values)	Nylon-66	Glass-Filled Nylon-66	Nylon-6	Glass-Filled Nylon-6
Heat deflection temperature @ 1820 kPa (°C)	75	250	80	210
Maximum resistance to continuous heat (°C)	120	140	125	130
Coefficient of linear expansion cm/cm/°C $\times 10^{-5}$	8.0	2.0	8.0	3.0
Compressive strength (kPa)	103,500	207,000	96,500	131,000
Flexural strength (kPa)	103,500	276,000	96,500	207,000
Impact strength (Izod: cm·N/cm of notch)	80	106.7	160	160
Tensile strength (kPa)	82,750	172,000	62,055	172,000
Elongation %	60	3	3	3
Hardness: Rockwell	R120	M100	M119	M101
Specific gravity	1.2	1.4	1.15	1.4

Table 13.13. Phenolic plastics

Thermal and Physical Properties of Plastics (Typical Values)	Wood-Flour-Filled	Mineral-Filled
Heat deflection temperature @ 1820 kPa (°C)	165	200
Maximum resistance to continuous heat (°C)	160	175
Coefficient of linear expansion cm/cm/°C $\times 10^{-5}$	3.0	2.0
Compressive strength (kPa)	172,400	172,400
Flexural strength (kPa)	62,000	82,750
Impact strength (Izod: cm·N/cm of notch)	21.5	21.5
Tensile strength (kPa)	48,250	41,400
Elongation %	0.5	0.5
Hardness: Rockwell	M100	M110
Specific gravity	1.4	1.5

Table 13.14. Poly(amide-imide)

Thermal and Physical Properties of Plastics (Typical Values)	Poly(Amide-Imide)
Heat deflection temperature @ 1820 kPa (°C)	275
Maximum resistance to continuous heat (°C)	225
Coefficient of linear expansion cm/cm/°C $\times 10^{-5}$	3.6
Compressive strength (kPa)	220,000
Flexural strength (kPa)	210,000
Impact strength (Izod: cm·N/cm of notch)	133.5
Tensile strength (kPa)	186,000
Elongation %	12
Hardness: Rockwell	M119
Specific gravity	1.4

Table 13.15. Polyaryl esters

Thermal and Physical Properties of Plastics (Typical Values)	Polyarylate	PET	PBT
Heat deflection temperature @ 1820 kPa (°C)	175	100	65
Maximum resistance to continuous heat (°C)	150	100	60
Coefficient of linear expansion cm/cm/°C $\times 10^{-5}$	6.5	6.5	7.0
Compressive strength (kPa)	93,000	86,000	75,800
Flexural strength (kPa)	79,800	112,300	96,500
Impact strength (Izod: cm·N/cm of notch)	215	26.7	53.4
Tensile strength (kPa)	68,000	62,000	55,000
Elongation %	50	100	100
Hardness: Rockwell	R125	M96	M70
Specific gravity	1.2	1.35	1.35

Table 13.16. Polycarbonates

Thermal and Physical Properties of Plastics (Typical Values)	Unfilled	20% Glass-Filled
Heat deflection temperature @ 1820 kPa (°C)	130	145
Maximum resistance to continuous heat (°C)	115	130
Coefficient of linear expansion cm/cm/°C $\times 10^{-5}$	6.8	2.2
Compressive strength (kPa)	86,000	124.000
Flexural strength (kPa)	93,000	158,600
Impact strength (Izod: cm·N/cm of notch)	534	106.7
Tensile strength (kPa)	72,000	131,000
Elongation %	110	4
Hardness: Rockwell	M70	M92
Specific gravity	1.2	1.4

Table 13.17. Polyesters (unsaturated)

Thermal and Physical Properties of Plastics (Typical Values)	Glass-Reinforced
Heat deflection temperature @ 1820 kPa (°C)	200
Maximum resistance to continuous heat (°C)	160
Coefficient of linear expansion cm/cm/°C $\times 10^{-5}$	2.5
Compressive strength (kPa)	172,000
Flexural strength (kPa)	83,000
Impact strength (Izod: cm·N/cm of notch)	160
Tensile strength (kPa)	69,000
Elongation %	1.5
Hardness: Rockwell	M50
Specific gravity	2

Table 13.18. Polyethylenes

Thermal and Physical Properties of Plastics (Typical Values)	LDPE	HDPE	UHMWPE	30% Glass-Filled HDPE
Heat deflection temperature @ 1820 kPa (°C)	40	85	85	120
Maximum resistance to continuous heat (°C)	40	80	80	110
Coefficient of linear expansion cm/cm/°C $\times 10^{-5}$	10.0	12.0	12.0	5.0
Compressive strength (kPa)	—	20,680	—	43,260
Flexural strength (kPa)	—	—	—	75,800
Impact strength (Izod: cm·N/cm of notch)	no break	106.7	no break	53.4
Tensile strength (kPa)	5,515	27,580	38,000	62,000
Elongation %	100	30	400	1.5
Hardness: Rockwell	D40	D40	R50	R75
Specific gravity	0.91	0.95	0.94	1.3

Table 13.19. Polyimides

Thermal and Physical Properties of Plastics (Typical Values)	Thermoplastic	50% Glass-Filled Thermoset
Heat deflection temperature @ 1820 kPa (°C)	315	350
Maximum resistance to continuous heat (°C)	300	325
Coefficient of linear expansion cm/cm/°C × 10^{-5}	5.0	1.3
Compressive strength (kPa)	241,300	234,400
Flexural strength (kPa)	172,400	144,800
Impact strength (Izod: cm·N/cm of notch)	80	294
Tensile strength (kPa)	96,500	44,000
Elongation %	8	0.5
Hardness: Rockwell	E60	M118
Specific gravity	1.4	1.6

Table 13.20. Polymethyl pentene

Thermal and Physical Properties of Plastics (Typical Values)	TPX
Heat deflection temperature @ 1820 kPa (°C)	80
Maximum resistance to continuous heat (°C)	40
Coefficient of linear expansion cm/cm/°C × 10^{-5}	11.7
Compressive strength (kPa)	38,000
Flexural strength (kPa)	34,500
Impact strength (Izod: cm·N/cm of notch)	27
Tensile strength (kPa)	24,130
Elongation %	15
Hardness: Rockwell	L70
Specific gravity	0.83

Table 13.21. Polyphenylene oxide

Thermal and Physical Properties of Plastics (Typical Values)	PPO	Glass-Filled PPO
Heat deflection temperature @ 1820 kPa (°C)	100	145
Maximum resistance to continuous heat (°C)	80	130
Coefficient of linear expansion cm/cm/°C $\times 10^{-5}$	5.0	2.0
Compressive strength (kPa)	96,500	123,000
Flexural strength (kPa)	89,600	144,800
Impact strength (Izod: cm · N/cm of notch)	270	107
Tensile strength (kPa)	55,150	120,500
Elongation %	50	4
Hardness: Rockwell	R115	R115
Specific gravity	1.1	1.1

Table 13.22. Polyphenylene sulfide

Thermal and Physical Properties of Plastics (Typical Values)	Unfilled PPS	40% Glass-Filled PPS
Heat deflection temperature @ 1820 kPa (°C)	135	250
Maximum resistance to continuous heat (°C)	110	200
Coefficient of linear expansion cm/cm/°C $\times 10^{-5}$	5.0	2.2
Compressive strength (kPa)	110,000	144,800
Flexural strength (kPa)	96,500	207,000
Impact strength (Izod: cm · N/cm of notch)	21	75
Tensile strength (kPa)	74,000	141,000
Elongation %	1.1	1
Hardness: Rockwell	R123	R123
Specific gravity	1.3	1.6

Table 13.23. Polypropylene

Thermal and Physical Properties of Plastics (Typical Values)	Unfilled PP	40% Talc-Filled PP
Heat deflection temperature @ 1820 kPa (°C)	55	100
Maximum resistance to continuous heat (°C)	100	120
Coefficient of linear expansion cm/cm/°C $\times 10^{-5}$	9.0	6.5
Compressive strength (kPa)	44,800	54,000
Flexural strength (kPa)	48,000	58,600
Impact strength (Izod: cm·N/cm of notch)	27	27
Tensile strength (kPa)	34,500	30,000
Elongation %	100	5
Hardness: Rockwell	R80	R95
Specific gravity	0.90	1.22

Table 13.24. Polystyrene

Thermal and Physical Properties of Plastics (Typical Values)	Unfilled PS	Impact PS	30% Glass-Filled PS
Heat deflection temperature @ 1820 kPa (°C)	90	90	105
Maximum resistance to continuous heat (°C)	75	70	95
Coefficient of linear expansion cm/cm/°C $\times 10^{-5}$	7.5	8.0	4.0
Compressive strength (kPa)	89,600	45,000	103,400
Flexural strength (kPa)	82,750	50,000	117,000
Impact strength (Izod: cm·N/cm of notch)	21	80	80
Tensile strength (kPa)	41,400	41,400	82,750
Elongation %	1.5	3	1
Hardness: Rockwell	M60	M35	M60
Specific gravity	1.04	1.04	1.2

Table 13.25. Polysulfones

Thermal and Physical Properties of Plastics (Typical Values)	Polysulfone	Polyphenyl Sulfone	Polyether Sulfone
Heat deflection temperature @ 1820 kPa (°C)	175	205	205
Maximum resistance to continuous heat (°C)	150	165	165
Coefficient of linear expansion cm/cm/°C $\times 10^{-5}$	5.4	3.1	5.5
Compressive strength (kPa)	96,500	86,000	96,500
Flexural strength (kPa)	107,000	86,000	127,000
Impact strength (Izod: cm · N/cm of notch)	80	640	80
Tensile strength (kPa)	82,750	82,750	82,750
Elongation %	25	25	25
Hardness: Rockwell	M69	M80	M88
Specific gravity	1.24	1.29	1.37

Table 13.26. Polyurethanes

Thermal and Physical Properties of Plastics (Typical Values)	Cast PU	Elastomer PU
Heat deflection temperature @ 1820 kPa (°C)	75	70
Maximum resistance to continuous heat (°C)	70	65
Coefficient of linear expansion cm/cm/°C $\times 10^{-5}$	15.0	15.0
Compressive strength (kPa)	138,000	138,000
Flexural strength (kPa)	20,680	27,500
Impact strength (Izod: cm · N/cm of notch)	1335.0	no break
Tensile strength (kPa)	6,890	7,250
Elongation %	200	150
Hardness: Rockwell	1.2	1.15
Specific gravity		

Table 13.27. Polyvinyl chloride

Thermal and Physical Properties of Plastics (Typical Values)	Rigid PVC	Flexible PVC	Chlorinated PVC
Heat deflection temperature @ 1820 kPa (°C)	65	—	105
Maximum resistance to continuous heat (°C)	60	35	90
Coefficient of linear expansion cm/cm/°C $\times 10^{-5}$	6.0	12.5	7.0
Compressive strength (kPa)	68,950	6,895	103,400
Flexural strength (kPa)	89,600	—	103,400
Impact strength (Izod: cm · N/cm of notch)	27		
Tensile strength (kPa)	44,800	10,340	55,150
Elongation %	50	200	5
Hardness: Rockwell	—	—	R120
Specific gravity	1.4	1.3	1.55

Table 13.28. Thermoplastic elastomers (TPE)

Thermal and Physical Properties of Plastics (Typical Values)	Styrene–Butadiene Block Copolymers	Polyolefin Blends	Polyester Block Copolymers
Heat deflection temperature @ 1820 kPa (°C)	25	25	30
Maximum resistance to continuous heat (°C)	50	50	50
Coefficient of linear expansion cm/cm/°C $\times 10^{-5}$	13.0	14.0	—
Compressive strength (kPa)	—	—	—
Flexural strength (kPa)	—	—	—
Impact strength (Izod: cm · N/cm of notch)	no break	no break	no break
Tensile strength (kPa)	6,895	6,895	27.580
Elongation %	600	150	400
Hardness: Rockwell	—	—	—
Specific gravity	1	0.9	1.2

Table 13.29. Silicones (SI)

Thermal and Physical Properties of Plastics (Typical Values)	Cast SI (RTV)	Glass-Filled-Molded SI
Heat deflection temperature @ 1820 kPa (°C)	—	260
Maximum resistance to continuous heat (°C)	—	260
Coefficient of linear expansion cm/cm/°C $\times 10^{-5}$	40.0	3.0
Compressive strength (kPa)	—	96,500
Flexural strength (kPa)	—	—
Impact strength (Izod: cm · N/cm of notch)	no break	53
Tensile strength (kPa)	3,450	34,450
Elongation %	200	4
Hardness: Rockwell	—	M80
Specific gravity	1.2	1.85

SUGGESTED READING

J. Agranoff, Ed., *Modern Plastics Encyclopedia,* McGraw-Hill, New York, 1980.

J. H. Dubois and F. W. John, *Plastics,* 5th ed., Van Nostrand-Reinhold Co., New York, 1974.

J. Frados, *SPI Plastics Engineering Handbook,* 4th ed., Van Nostrand-Reinhold Co., New York, 1976.

C. A. Harper, Ed., *Handbook of Plastics and Elastomers,* McGraw-Hill, New York, 1975.

D. C. Miles and J. H. Briston, *Polymer Technology,* Chemical Publishing Co., New York, 1979.

R. V. Milby, *Plastics Technology,* McGraw-Hill, New York, 1973.

R. B. Seymour, *Modern Plastics Technology,* Reston Publishing Co., Reston, VA, 1975.

CHAPTER 14

Selection Guide: Chemical Resistance Data for Typical Commercial Plastics

As pointed out in previous chapters, plastic materials for potential applications as corrosion resistant materials applications can be screened by use of a knowledge of their molecular structure and physical and thermal properties.

The general chemical resistance data for typical corrosive environments shown in the first part of the following tables are also very helpful in predicting performance of plastics in corrosive environments. In general, most water-insoluble plastics are resistant to aqueous salts and alkalies. Many are also resistant to nonoxidizing acids, such as 25% sulfuric acid and polar solvents such as ethanol. Fewer plastics are resistant to oxidizing acids such as 20% nitric acid and nonpolar solvents such as benzene.

The relative resistance of typical plastics to many specific corrosive solutions is also shown in the following tables. Although specific plastic materials can be selected by use of these data, the final decision should be based on simulated tests, case history data, and/or recommendations from reliable suppliers.

Considerable information may also be obtained from the suggested reading at the end of this chapter.

Corrosion resistant data are based on tests with typical generic plastics. Trade names for some of these corresponding commercial products are also shown in a table at the end of this chapter.

Table 14.1. Chemical resistance data for plastics[a,b]

Corrosive	25°C	65°C	90°C
		ABS Terpolymer	
Nonoxidizing acids (20% H_2SO_4)	S	S	U
Oxidizing acids (10% HNO_3)	Q	U	U
Aqueous salt solution (NaCl)	S	S	U
Aqueous alkalies (NaOH)	S	S	U
Polar solvents (C_2H_5OH)	Q	U	U
Nonpolar solvents (C_6H_6)	U	U	U
Water	S	S	U
Acetaldehyde	U	U	U
Acetic acid (glacial)	U	U	U
Acetic acid (1–50%)	S	U	U
Acetic anhydride	U	U	U
Acetone	U	U	U
Acetonitrile	U	U	U
Acetophenone	U	U	U
Acetyl chloride	U	U	U
Acrylonitrile	U	U	U
Adipic acid	S	S	U
Allyl alcohol	U	U	U
Allyl chloride	U	U	U
Alum (aqueous solutions)	S	S	U
Aluminum salts	S	S	U
Ammonia (anhydrous)	U	U	U
Ammonia (aqueous)	S	S	U
Ammonium salts	S	S	U
Amyl acetate	U	U	U
Amyl alcohol	S	Q	U
Amyl chloride	U	U	U
Aniline	U	U	U
Antimony salts	S	S	U
Aqua regia	U	U	U
Arsenic salts	S	S	U
Barium salts	S	S	U
Barium hydroxide	S	S	U
Beer	S	S	U
Benzaldehyde	U	U	U
Benzene	U	U	U
Benzene sulfonic acid	S	S	U
Benzoic acid	S	S	U
Benzyl alcohol	U	U	U
Benzyl chloride	U	U	U
Bismuth salts	S	S	U
Borax	S	S	U

Table 14.1. (Continued)

Corrosive	25°C	65°C	90°C
Boric acid	S	S	U
Bromine	U	U	U
Butadiene	Q	U	U
Butane	Q	U	U
Butyl acetate	U	U	U
Butyl alcohol	U	U	U
Butylamine	U	U	U
Butyl bromide	U	U	U
Butyl chloride	U	U	U
Calcium salts	S	S	U
Calcium hydroxide	S	S	U
Calcium hypochlorite	S	Q	U
Carbon disulfide	U	U	U
Carbon tetrachloride	U	U	U
Carbonic acid (aq. CO_2)	S	S	U
Castor oil	S	Q	U
Caustic (aqueous)	S	S	U
Cellosolve	U	U	U
Chlorine	U	U	U
Chlorine dioxide	U	U	U
Chloroacetic acid	U	U	U
Chlorobenzene	U	U	U
Chloroform	U	U	U
Chromic acid (40%)	U	U	U
Chromic acid (10%)	S	U	U
Citric acid	S	S	U
Clorox bleach	S	S	U
Copper salts	S	S	U
Cresol	U	U	U
Cyclohexane	Q	U	U
Cyclohexanol	Q	U	U
Cyclohexanone	U	U	U
Detergents (general)	S	U	U
Dibutyl phthalate	U	U	U
Diesel fuel	Q	U	U
Diethylamine	U	U	U
Dimethylformamide	U	U	U
Dioctyl phthalate	U	U	U
Dioxane	U	U	U
Esters (general)	U	U	U
Ethyl acetate	U	U	U
Ethanol (1–95%)	S	Q	U
Ethyl chloride	U	U	U
Ethyl ether	U	U	U

Table 14.1. (Continued)

Corrosive	25°C	65°C	90°C
Ethylene chloride	U	U	U
Ethylenediamine	U	U	U
Ethylene glycol	S	S	U
Fatty acids (general)	S	Q	U
Fluorine	U	U	U
Formaldehyde (37%)	S	S	U
Formic acid	U	U	U
Freons	Q	U	U
Fructose	S	S	U
Furfural	U	U	U
Gasohol	Q	U	U
Gasolene	Q	U	U
Glucose	S	S	U
Glycerol	S	S	U
Heptane	Q	S	U
Hexane	Q	Q	U
Hexanol	U	U	U
Hydrobromic acid	S	Q	U
Hydrochloric acid (35%)	S	Q	U
Hydrochloric acid (1–25%)	S	S	U
Hydrofluoric acid	U	U	U
Hydrogen peroxide (3–10%)	U	U	U
Hydrogen sulfide	S	Q	U
Iodine	Q	Q	U
Iron salts	S	S	U
Isopropyl alcohol	S	Q	U
Kerosene	S	Q	U
Lead salts	S	S	U
Linseed oil	S	Q	U
Lithium salts	S	S	U
Magnesium salts	S	S	U
Magnesium hydroxide	S	S	U
Mercury salts	S	S	S
Methanol	Q	U	U
Methyl ethyl ketone	Q	Q	U
Methyl isobutyl ketone	U	U	U
Methyl sulfate	S	S	U
Milk	S	S	U
Mineral oil	S	Q	U
Naphtha	S	Q	U
Nickel salts	S	S	U
Nitric acid (conc.)	U	U	U
Nitric acid (20%)	Q	U	U
Nitric acid (10%)	S	Q	U

Table 14.1. (Continued)

Corrosive	25°C	65°C	90°C
Nitrobenzene	U	U	U
Octane	Q	Q	U
Oleum	U	U	U
Oxalic acid	S	S	U
Perchloroethylene	U	U	U
Perchloric acid	U	U	U
Phenol	U	U	U
Phosphoric acid	S	Q	U
Phthalic acid	S	S	U
Phthalic anhydride	S	S	U
Potassium hydroxide	S	S	U
Potassium permanganate	S	S	U
Potassium salts	S	S	U
Pyridine	Q	U	U
Silver salts	U	U	U
Soap	S	S	U
Sodium salts	S	S	U
Sodium chlorate	Q	S	U
Sodium hydroxide	S	S	U
Sodium hypochlorite	S	S	U
Sulfur dioxide	S	Q	U
Sulfur trioxide	U	U	U
Sulfuric acid (100%)	U	U	U
Sulfuric acid (70%)	S	Q	U
Sulfuric acid (1–50%)	S	S	U
Tetralin	U	U	U
Tin salts	S	S	U
Toluene	U	U	U
Trichloroethylene	U	U	U
Turpentine	U	U	U
Urine	S	S	U
Vegetable oils	S	Q	U
Wine	S	S	U
Xylene	U	U	U
Zinc salts	S	S	U
		Acetal	
Nonoxidizing acids (20% H_2SO_4)	U	U	U
Oxidizing acids (10% HNO_3)	U	U	U
Aqueous salt solution (NaCl)	S	S	U
Aqueous alkalies (NaOH)	S	Q	U
Polar solvents (C_2H_5OH)	S	Q	U
Nonpolar solvents (C_6H_6)	Q	U	U
Water	S	S	U

Table 14.1. (Continued)

Corrosive	25°C	65°C	90°C
Acetaldehyde	S	Q	U
Acetic acid (glacial)	U	U	U
Acetic acid (1–50%)	Q	U	U
Acetic anhydride	U	U	U
Acetone	U	U	U
Acetonitrile	U	U	U
Acetophenone	U	U	U
Acetyl chloride	U	U	U
Acrylonitrile	U	U	U
Adipic acid	S	Q	U
Allyl alcohol	S	Q	U
Allyl chloride	S	U	U
Alum (aqueous solutions)	S	S	U
Aluminum salts	S	S	U
Ammonia (anhydrous)	S	S	U
Ammonia (aqueous)	S	Q	U
Ammonium salts	S	S	U
Amyl acetate	S	Q	U
Amyl alcohol	S	Q	U
Amyl chloride	S	Q	U
Aniline	Q	U	U
Antimony salts	S	S	U
Aqua regia	U	U	U
Arsenic salts	S	S	U
Barium salts	S	S	U
Barium hydroxide	S	S	U
Beer	S	S	U
Benzaldehyde	Q	U	U
Benzene	Q	U	U
Benzene sulfonic acid	S	U	U
Benzoic acid	S	U	U
Benzyl alcohol	Q	U	U
Benzyl chloride	Q	U	U
Bismuth salts	S	Q	U
Borax	S	U	U
Boric acid	S	U	U
Bromine	U	U	U
Butadiene	S	Q	U
Butane	S	S	U
Butyl acetate	S	U	U
Butyl alcohol	S	Q	U
Butylamine	S	Q	U
Butyl bromide	S	U	U
Butyl chloride	S	U	U

Table 14.1. (Continued)

Corrosive	25°C	65°C	90°C
Calcium salts	S	S	U
Calcium hydroxide	S	S	U
Calcium hypochlorite	Q	Q	U
Carbon disulfide	S	U	U
Carbon tetrachloride	S	Q	U
Carbonic acid (aq. CO_2)	S	S	U
Castor oil	S	Q	U
Caustic (aqueous)	S	S	U
Cellosolve	S	Q	U
Chlorine	U	U	U
Chlorine dioxide	U	U	U
Chloroacetic acid	U	U	U
Chlorobenzene	S	U	U
Chloroform	S	Q	U
Chromic acid (40%)	U	U	U
Chromic acid (10%)	U	U	U
Citric acid	Q	U	U
Clorox bleach	S	Q	U
Copper salts	S	Q	U
Cresol	Q	U	U
Cyclohexane	S	U	U
Cyclohexanol	S	U	U
Cyclohexanone	S	Q	U
Detergents (general)	Q	Q	U
Dibutyl phthalate	S	S	U
Diesel fuel	S	Q	U
Diethylamine	U	U	U
Dimethylformamide	U	U	U
Dioctyl phthalate	Q	U	U
Dioxane	S	Q	U
Esters (general)	S	U	U
Ethyl acetate	S	U	U
Ethanol (1–95%)	S	S	U
Ethyl chloride	Q	U	U
Ethyl ether	Q	U	U
Ethylene chloride	S	U	U
Ethylenediamine	S	Q	U
Ethylene glycol	S	S	U
Fatty acids (general)	Q	U	U
Fluorine	U	U	U
Formaldehyde (37%)	S	S	U
Formic acid	U	U	U
Freons	S	U	U
Fructose	S	S	U

Table 14.1. (Continued)

Corrosive	25°C	65°C	90°C
Furfural	Q	U	U
Gasohol	S	S	U
Gasolene	S	Q	U
Glucose	S	S	U
Glycerol	S	S	U
Heptane	S	S	U
Hexane	S	S	U
Hexanol	S	S	U
Hydrobromic acid	U	U	U
Hydrochloric acid (35%)	U	U	U
Hydrochloric acid (1–25%)	U	U	U
Hydrofluoric acid	U	U	U
Hydrogen peroxide (3–10%)	Q	U	U
Hydrogen sulfide	S	U	U
Iodine	S	Q	U
Iron salts	S	S	U
Isopropyl alcohol	S	S	U
Kerosene	S	S	U
Lead salts	S	S	U
Linseed oil	S	S	U
Lithium salts	S	S	U
Magnesium salts	S	S	U
Magnesium hydroxide	S	S	U
Mercury salts	S	S	U
Methanol	S	S	U
Methyl ethyl ketone	Q	U	U
Methyl isobutyl ketone	Q	U	U
Methyl sulfate	U	U	U
Milk	S	S	U
Mineral oil	S	Q	U
Naphtha	S	Q	U
Nickel salts	S	S	U
Nitric acid (conc.)	U	U	U
Nitric acid (20%)	U	U	U
Nitric acid (10%)	U	U	U
Nitrobenzene	S	Q	U
Octane	S	S	U
Oleum	U	U	U
Oxalic acid	Q	U	U
Perchloroethylene	Q	U	U
Perchloric acid	U	U	U
Phenol	U	U	U
Phosphoric acid	U	U	U
Phthalic acid	Q	U	U

Table 14.1. (Continued)

Corrosive	25°C	65°C	90°C
Phthalic anhydride	Q	U	U
Potassium hydroxide	S	S	U
Potassium permanganate	Q	U	U
Potassium salts	S	S	U
Pyridine	S	S	U
Silver salts	S	S	U
Soap	S	S	U
Sodium salts	S	S	U
Sodium chlorate	U	U	U
Sodium hydroxide	S	S	U
Sodium hypochlorite	Q	U	U
Sulfur dioxide	Q	U	U
Sulfur trioxide	U	U	U
Sulfuric acid (100%)	U	U	U
Sulfuric acid (70%)	U	U	U
Sulfuric acid (1–50%)	U	U	U
Tetralin	S	S	U
Tin salts	S	S	U
Toluene	Q	U	U
Trichloroethylene	Q	U	U
Turpentine	S	S	U
Urine	S	S	U
Vegetable oils	S	S	U
Wine	S	Q	U
Xylene	S	Q	U
Zinc salts	S	S	U
		Acrylics (PMMA)	
Nonoxidizing acids (20% H_2SO_4)	S	Q	U
Oxidizing acids (10% HNO_3)	U	U	U
Aqueous salt solution (NaCl)	S	S	U
Aqueous alkalies (NaOH)	S	Q	U
Polar solvents (C_2H_5OH)	S	Q	U
Nonpolar solvents (C_6H_6)	U	U	U
Water	S	S	U
Acetaldehyde	U	U	U
Acetic acid (glacial)	U	U	U
Acetic acid (1–50%)	S	Q	U
Acetic anhydride	U	U	U
Acetone	U	U	U
Acetonitrile	U	U	U
Acetophenone	U	U	U
Acetyl chloride	U	U	U

Table 14.1. (Continued)

Corrosive	25°C	65°C	90°C
Acrylonitrile	S	Q	U
Adipic acid	S	Q	U
Allyl alcohol	Q	U	U
Allyl chloride	U	U	U
Alum (aqueous solutions)	S	S	U
Aluminum salts	S	S	U
Ammonia (anhydrous)	Q	U	U
Ammonia (aqueous)	Q	U	U
Ammonium salts	S	S	U
Amyl acetate	U	U	U
Amyl alcohol	U	U	U
Amyl chloride	U	U	U
Aniline	U	U	U
Antimony salts	S	S	U
Aqua regia	U	U	U
Arsenic salts	S	S	U
Barium salts	S	S	U
Barium hydroxide	S	S	U
Beer	S	S	U
Benzaldehyde	U	U	U
Benzene	U	U	U
Benzene sulfonic acid	Q	U	U
Benzoic acid	S	S	U
Benzyl alcohol	U	U	U
Benzyl chloride	U	U	U
Bismuth salts	S	S	U
Borax	S	S	U
Boric acid	S	S	U
Bromine	S	S	U
Butadiene	U	U	U
Butane	U	U	U
Butyl acetate	U	U	U
Butyl alcohol	U	U	U
Butylamine	U	U	U
Butyl bromide	U	U	U
Butyl chloride	U	U	U
Calcium salts	S	S	U
Calcium hydroxide	S	S	U
Calcium hypochlorite	S	S	U
Carbon disulfide	U	U	U
Carbon tetrachloride	U	U	U
Carbonic acid (aq. CO_2)	S	S	U
Castor oil	S	Q	U
Caustic (aqueous)	S	Q	U

Table 14.1. (Continued)

Corrosive	25°C	65°C	90°C
Cellosolve	U	U	U
Chlorine	U	U	U
Chlorine dioxide	U	U	U
Chloroacetic acid	U	U	U
Chlorobenzene	U	U	U
Chloroform	U	U	U
Chromic acid (40%)	U	U	U
Chromic acid (10%)	U	U	U
Citric acid	S	S	U
Clorox bleach	Q	U	U
Copper salts	S	S	U
Cresol	U	U	U
Cyclohexane	S	Q	U
Cyclohexanol	S	U	U
Cyclohexanone	U	U	U
Detergents (general)	S	Q	U
Dibutyl phthalate	U	U	U
Diesel fuel	S	Q	U
Diethylamine	U	U	U
Dimethylformamide	U	U	U
Dioctyl phthalate	U	U	U
Dioxane	U	U	U
Esters (general)	U	U	U
Ethyl acetate	U	U	U
Ethanol (1–95%)	S	U	U
Ethyl chloride	U	U	U
Ethyl ether	U	U	U
Ethylene chloride	U	U	U
Ethylenediamine	U	U	U
Ethylene glycol	S	S	U
Fatty acids (general)	S	Q	U
Fluorine	U	U	U
Formaldehyde (37%)	S	Q	U
Formic acid	S	Q	U
Freons	Q	U	U
Fructose	S	S	U
Furfural	U	U	U
Gasohol	S	Q	U
Gasolene	S	Q	U
Glucose	S	S	U
Glycerol	S	S	U
Heptane	S	Q	U
Hexane	S	Q	U
Hexanol	S	Q	U

Table 14.1. (Continued)

Corrosive	25°C	65°C	90°C
Hydrobromic acid	S	Q	U
Hydrochloric acid (35%)	S	Q	U
Hydrochloric acid (1–25%)	S	Q	U
Hydrofluoric acid	U	U	U
Hydrogen peroxide (3–10%)	Q	U	U
Hydrogen sulfide	S	Q	U
Iodine	S	Q	U
Iron salts	S	S	U
Isopropyl alcohol	S	U	U
Kerosene	S	Q	U
Lead salts	S	S	U
Linseed oil	Q	U	U
Lithium salts	S	S	U
Magnesium salts	S	S	U
Magnesium hydroxide	S	S	U
Mercury salts	S	S	U
Methanol	Q	U	U
Methyl ethyl ketone	U	U	U
Methyl isobutyl ketone	U	U	U
Methyl sulfate	U	U	U
Milk	S	S	U
Mineral oil	S	Q	U
Naphtha	S	U	U
Nickel salts	S	S	U
Nitric acid (conc.)	U	U	U
Nitric acid (20%)	U	U	U
Nitric acid (10%)	U	U	U
Nitrobenzene	U	U	U
Octane	S	Q	U
Oleum	U	U	U
Oxalic acid	S	Q	U
Perchloroethylene	U	U	U
Perchloric acid	U	U	U
Phenol	Q	U	U
Phosphoric acid	Q	U	U
Phthalic acid	S	Q	U
Phthalic anhydride	S	S	U
Potassium hydroxide	S	Q	U
Potassium permanganate	Q	U	U
Potassium salts	S	S	U
Pyridine	U	U	U
Silver salts	S	S	U
Soap	S	S	U
Sodium salts	S	S	U

Table 14.1. (Continued)

Corrosive	25°C	65°C	90°C
Sodium chlorate	S	Q	U
Sodium hydroxide	S	Q	U
Sodium hypochlorite	S	Q	U
Sulfur dioxide	S	Q	U
Sulfur trioxide	U	U	U
Sulfuric acid (100%)	U	U	U
Sulfuric acid (70%)	Q	U	U
Sulfuric acid (1–50%)	Q	U	U
Tetralin	U	U	U
Tin salts	S	S	Q
Toluene	U	U	U
Trichloroethylene	U	U	U
Turpentine	U	U	U
Urine	S	S	U
Vegetable oils	S	Q	U
Wine	S	S	U
Xylene	U	U	U
Zinc salts	S	S	U
		Alkyds	
Nonoxidizing acids (20% H_2SO_4)	S	U	U
Oxidizing acids (10% HNO_3)	U	U	U
Aqueous salt solution (NaCl)	S	S	U
Aqueous alkalies (NaOH)	Q	U	U
Polar solvents (C_2H_5OH)	S	S	U
Nonpolar solvents (C_6H_6)	U	U	U
Water	S	S	Q
Acetaldehyde	U	U	U
Acetic acid (glacial)	Q	U	U
Acetic acid (1–50%)	S	U	U
Acetic anhydride	U	U	U
Acetone	U	U	U
Acetonitrile	U	U	U
Acetophenone	U	U	U
Acetyl chloride	U	U	U
Acrylonitrile	U	U	U
Adipic acid	S	U	U
Allyl alcohol	Q	U	U
Allyl chloride	U	U	U
Alum (aqueous solutions)	S	S	U
Aluminum salts	S	S	U
Ammonia (anhydrous)	Q	U	U
Ammonia (aqueous)	Q	U	U

Table 14.1. (Continued)

Corrosive	25°C	65°C	90°C
Ammonium salts	S	S	U
Amyl acetate	U	U	U
Amyl alcohol	S	U	U
Amyl chloride	U	U	U
Aniline	U	U	U
Antimony salts	S	S	U
Aqua regia	U	U	U
Arsenic salts	S	S	U
Barium salts	S	S	U
Barium hydroxide	S	Q	U
Beer	S	S	U
Benzaldehyde	U	U	U
Benzene	Q	U	U
Benzene sulfonic acid	Q	U	U
Benzoic acid	S	U	U
Benzyl alcohol	Q	U	U
Benzyl chloride	U	U	U
Bismuth salts	S	S	U
Borax	Q	Q	U
Boric acid	S	U	U
Bromine	U	U	U
Butadiene	U	U	U
Butane	Q	U	U
Butyl acetate	U	U	U
Butyl alcohol	S	U	U
Butylamine	U	U	U
Butyl bromide	U	U	U
Butyl chloride	U	U	U
Calcium salts	S	S	U
Calcium hydroxide	S	Q	U
Calcium hypochlorite	Q	U	U
Carbon disulfide	U	U	U
Carbon tetrachloride	S	U	U
Carbonic acid (aq. CO_2)	S	S	U
Castor oil	S	U	U
Caustic (aqueous)	Q	U	U
Cellosolve	U	U	U
Chlorine	U	U	U
Chlorine dioxide	Q	U	U
Chloroacetic acid	U	U	U
Chlorobenzene	U	U	U
Chloroform	Q	U	U
Chromic acid (40%)	U	U	U
Chromic acid (10%)	S	U	U

Table 14.1. (Continued)

Corrosive	25°C	65°C	90°C
Citric acid	S	S	U
Clorox bleach	Q	U	U
Copper salts	S	S	U
Cresol	U	U	U
Cyclohexane	S	Q	U
Cyclohexanol	S	Q	U
Cyclohexanone	U	U	U
Detergents (general)	S	Q	U
Dibutyl phthalate	U	U	U
Diesel fuel	S	Q	U
Diethylamine	U	U	U
Dimethylformamide	U	U	U
Dioctyl phthalate	U	U	U
Dioxane	U	U	U
Esters (general)	Q	U	U
Ethyl acetate	Q	U	U
Ethanol (1–95%)	S	Q	U
Ethyl chloride	U	U	U
Ethyl ether	U	U	U
Ethylene chloride	U	U	U
Ethylenediamine	Q	U	U
Ethylene glycol	S	S	U
Fatty acids (general)	S	S	U
Fluorine	U	U	U
Formaldehyde (37%)	S	Q	U
Formic acid	Q	U	U
Freons	Q	U	U
Fructose	S	S	U
Furfural	U	U	U
Gasohol	S	U	U
Gasolene	S	U	U
Glucose	S	S	U
Glycerol	S	S	U
Heptane	S	U	U
Hexane	S	U	U
Hexanol	S	U	U
Hydrobromic acid	Q	U	U
Hydrochloric acid (35%)	Q	U	U
Hydrochloric acid (1–25%)	Q	U	U
Hydrofluoric acid	U	U	U
Hydrogen peroxide (3–10%)	S	U	U
Hydrogen sulfide	S	Q	U
Iodine	Q	U	U
Iron salts	S	S	U

Table 14.1. (Continued)

Corrosive	25°C	65°C	90°C
Isopropyl alcohol	S	S	U
Kerosene	S	Q	U
Lead salts	S	S	U
Linseed oil	S	U	U
Lithium salts	S	S	U
Magnesium salts	S	S	U
Magnesium hydroxide	S	S	U
Mercury salts	S	S	U
Methanol	S	Q	U
Methyl ethyl ketone	Q	U	U
Methyl isobutyl ketone	Q	U	U
Methyl sulfate	Q	U	U
Milk	S	S	U
Mineral oil	S	Q	U
Naphtha	Q	U	U
Nickel salts	S	S	U
Nitric acid (conc.)	U	U	U
Nitric acid (20%)	U	U	U
Nitric acid (10%)	Q	U	U
Nitrobenzene	U	U	U
Octane	S	Q	U
Oleum	U	U	U
Oxalic acid	S	Q	U
Perchloroethylene	Q	U	U
Perchloric acid	Q	U	U
Phenol	U	U	U
Phosphoric acid	S	Q	U
Phthalic acid	S	S	U
Phthalic anhydride	S	Q	U
Potassium hydroxide	Q	U	U
Potassium permanganate	S	Q	U
Potassium salts	S	S	U
Pyridine	U	U	U
Silver salts	S	S	U
Soap	S	Q	U
Sodium salts	S	S	U
Sodium chlorate	Q	U	U
Sodium hydroxide	Q	U	U
Sodium hypochlorite	Q	U	U
Sulfur dioxide	S	U	U
Sulfur trioxide	S	U	U
Sulfuric acid (100%)	U	U	U
Sulfuric acid (70%)	U	U	U
Sulfuric acid (1–50%)	Q	U	U

Table 14.1. (Continued)

Corrosive	25°C	65°C	90°C
Tetralin	Q	U	U
Tin salts	S	S	U
Toluene	Q	U	U
Trichloroethylene	U	U	U
Turpentine	Q	U	U
Urine	S	S	U
Vegetable oils	S	Q	S
Wine	S	S	U
Xylene	Q	S	S
Zinc salts	S	S	U
		Allylic	
Nonoxidizing acids (20% H_2SO_4)	S	Q	U
Oxidizing acids (10% HNO_3)	U	U	U
Aqueous salt solution (NaCl)	S	S	Q
Aqueous alkalies (NaOH)	Q	U	U
Polar solvents (C_2H_5OH)	S	S	U
Nonpolar solvents (C_6H_6)	U	U	U
Water	S	S	Q
Acetaldehyde	U	U	U
Acetic acid (glacial)	U	U	U
Acetic acid (1–50%)	S	U	U
Acetic anhydride	U	U	U
Acetone	U	U	U
Acetonitrile	U	U	U
Acetophenone	U	U	U
Acetyl chloride	U	U	U
Acrylonitrile	U	U	U
Adipic acid	S	S	U
Allyl alcohol	Q	Q	U
Allyl chloride	U	U	U
Alum (aqueous solutions)	S	S	U
Aluminum salts	S	S	U
Ammonia (anhydrous)	U	U	U
Ammonia (aqueous)	S	Q	U
Ammonium salts	S	S	U
Amyl acetate	U	U	U
Amyl alcohol	S	Q	U
Amyl chloride	U	U	U
Aniline	U	U	U
Antimony salts	S	S	U
Aqua regia	U	U	U
Arsenic salts	S	S	U

Table 14.1. (Continued)

Corrosive	25°C	65°C	90°C
Barium salts	S	S	U
Barium hydroxide	S	Q	U
Beer	S	S	U
Benzaldehyde	U	U	U
Benzene	U	U	U
Benzene sulfonic acid	Q	U	U
Benzoic acid	S	Q	U
Benzyl alcohol	U	U	U
Benzyl chloride	U	U	U
Bismuth salts	S	S	U
Borax	S	Q	U
Boric acid	S	S	U
Bromine	U	U	U
Butadiene	Q	U	U
Butane	S	U	U
Butyl acetate	U	U	U
Butyl alcohol	S	Q	U
Butylamine	Q	U	U
Butyl bromide	U	U	U
Butyl chloride	U	U	U
Calcium salts	S	S	U
Calcium hydroxide	S	U	U
Calcium hypochlorite	Q	U	U
Carbon disulfide	U	U	U
Carbon tetrachloride	Q	U	U
Carbonic acid (aq. CO_2)	S	S	U
Castor oil			
Caustic (aqueous)	Q	U	U
Cellosolve	Q	U	U
Chlorine	Q	U	U
Chlorine dioxide	Q	U	U
Chloroacetic acid	U	U	U
Chlorobenzene	U	U	U
Chloroform	Q	U	U
Chromic acid (40%)	U	U	U
Chromic acid (10%)	U	U	U
Citric acid	S	S	Q
Clorox bleach	Q	U	U
Copper salts	S	S	Q
Cresol	S	S	Q
Cyclohexane	U	U	U
Cyclohexanol	U	U	U
Cyclohexanone	U	U	U
Detergents (general)	U	U	U

Table 14.1. (Continued)

Corrosive	25°C	65°C	90°C
Dibutyl phthalate	U	U	U
Diesel fuel	S	U	U
Diethylamine	S	Q	U
Dimethylformamide	U	U	U
Dioctyl phthalate	U	U	U
Dioxane	U	U	U
Esters (general)	U	U	U
Ethyl acetate	U	U	U
Ethanol (1–95%)	U	U	U
Ethyl chloride	U	U	U
Ethyl ether	U	U	U
Ethylene chloride	U	U	U
Ethylenediamine	U	U	U
Ethylene glycol	S	S	U
Fatty acids (general)	S	Q	U
Fluorine	U	U	U
Formaldehyde (37%)	S	S	U
Formic acid	Q	U	U
Freons	S	U	U
Fructose	S	S	Q
Furfural	U	U	U
Gasohol	S	Q	U
Gasolene	S	Q	U
Glucose	S	S	Q
Glycerol	S	S	Q
Heptane	S	Q	U
Hexane	S	Q	U
Hexanol	S	Q	U
Hydrobromic acid	S	Q	Q
Hydrochloric acid (35%)	S	Q	U
Hydrochloric acid (1–25%)	S	S	U
Hydrofluoric acid	S	Q	U
Hydrogen peroxide (3–10%)	S	U	U
Hydrogen sulfide	S	Q	U
Iodine	S	Q	U
Iron salts	S	S	U
Isopropyl alcohol	S	Q	U
Kerosene	S	S	U
Lead salts	S	S	U
Linseed oil	S	Q	U
Lithium salts	S	S	U
Magnesium salts	S	S	U
Magnesium hydroxide	S	S	U
Mercury salts	S	S	Q

Table 14.1. (Continued)

Corrosive	25°C	65°C	90°C
Methanol	S	Q	U
Methyl ethyl ketone	U	U	U
Methyl isobutyl ketone	U	U	U
Methyl sulfate	Q	U	U
Milk	S	S	U
Mineral oil	S	Q	U
Naphtha	S	Q	U
Nickel salts	S	S	Q
Nitric acid (conc.)	U	U	U
Nitric acid (20%)	U	U	U
Nitric acid (10%)	Q	U	U
Nitrobenzene	U	U	U
Octane	S	Q	U
Oleum	U	U	U
Oxalic acid	S	S	Q
Perchloroethylene	U	U	U
Perchloric acid	U	U	U
Phenol	U	U	U
Phosphoric acid	S	S	U
Phthalic acid	S	S	Q
Phthalic anhydride	S	S	U
Potassium hydroxide	U	U	U
Potassium permanganate	S	Q	U
Potassium salts	S	S	Q
Pyridine	U	U	U
Silver salts	S	S	U
Soap	S	Q	U
Sodium salts	S	S	U
Sodium chlorate	Q	U	U
Sodium hydroxide	U	U	U
Sodium hypochlorite	Q	U	U
Sulfur dioxide	S	S	U
Sulfur trioxide	S	Q	U
Sulfuric acid (100%)	U	U	U
Sulfuric acid (70%)	Q	U	U
Sulfuric acid (1–50%)	S	S	Q
Tetralin	Q	U	U
Tin salts	S	S	Q
Toluene	U	U	U
Trichloroethylene	U	U	U
Turpentine	U	U	U
Urine	S	S	Q
Vegetable oils	S	Q	U
Wine	S	S	U

Table 14.1. (Continued)

Corrosive	25°C	65°C	90°C
Xylene	U	U	U
Zinc salts	S	S	Q
		Amino	
Nonoxidizing acids (20% H_2SO_4)	Q	U	U
Oxidizing acids (10% HNO_3)	U	U	U
Aqueous salt solution (NaCl)	S	Q	U
Aqueous alkalies (NaOH)	Q	U	U
Polar solvents (C_2H_5OH)	S	U	U
Nonpolar solvents (C_6H_6)	Q	U	U
Water	S	Q	U
Acetaldehyde	U	U	U
Acetic acid (glacial)	U	U	U
Acetic acid (1–50%)	Q	U	U
Acetic anhydride	U	U	U
Acetone	U	U	U
Acetonitrile	U	U	U
Acetophenone	U	U	U
Acetyl chloride	U	U	U
Acrylonitrile	U	U	U
Adipic acid	S	Q	U
Allyl alcohol	Q	U	U
Allyl chloride	U	U	U
Alum (aqueous solutions)	S	U	U
Aluminum salts	S	U	U
Ammonia (anhydrous)	Q	U	U
Ammonia (aqueous)	Q	U	U
Ammonium salts	S	U	U
Amyl acetate	U	U	U
Amyl alcohol	Q	U	U
Amyl chloride	U	U	U
Aniline	U	U	U
Antimony salts	S	Q	U
Aqua regia	U	U	U
Arsenic salts	S	Q	U
Barium salts	S	Q	U
Barium hydroxide	S	U	U
Beer	S	Q	U
Benzaldehyde	U	U	U
Benzene	U	U	U
Benzene sulfonic acid	U	U	U
Benzoic acid	Q	U	U
Benzyl alcohol	U	U	U

Table 14.1. (Continued)

Corrosive	25°C	65°C	90°C
Benzyl chloride	U	U	U
Bismuth salts	S	Q	U
Borax	S	Q	U
Boric acid	U	U	U
Bromine	U	U	U
Butadiene	Q	U	U
Butane	Q	U	U
Butyl acetate	U	U	U
Butyl alcohol	U	U	U
Butylamine	U	U	U
Butyl bromide	U	U	U
Butyl chloride	U	U	U
Calcium salts	S	Q	U
Calcium hydroxide	S	Q	U
Calcium hypochlorite	Q	U	U
Carbon disulfide	U	U	U
Carbon tetrachloride	Q	U	U
Carbonic acid (aq. CO_2)	S	Q	U
Castor oil	S	U	U
Caustic (aqueous)	Q	U	U
Cellosolve	U	U	U
Chlorine	U	U	U
Chlorine dioxide	U	U	U
Chloroacetic acid	U	U	U
Chlorobenzene	U	U	U
Chloroform	Q	U	U
Chromic acid (40%)	U	U	U
Chromic acid (10%)	U	U	U
Citric acid	S	Q	U
Clorox bleach	Q	U	U
Copper salts	S	Q	U
Cresol	U	U	U
Cyclohexane	Q	U	U
Cyclohexanol	Q	U	U
Cyclohexanone	U	U	U
Detergents (general)	S	Q	U
Dibutyl phthalate	U	U	U
Diesel fuel	S	U	U
Diethylamine	U	U	U
Dimethylformamide	U	U	U
Dioctyl phthalate	U	U	U
Dioxane	U	U	U
Esters (general)	U	U	U
Ethyl acetate	U	U	U

Table 14.1. (Continued)

Corrosive	25°C	65°C	90°C
Ethanol (1–95%)	S	U	U
Ethyl chloride	U	U	U
Ethyl ether	U	U	U
Ethylene chloride	U	U	U
Ethylenediamine	U	U	U
Ethylene glycol	S	Q	U
Fatty acids (general)	S	Q	U
Fluorine	U	U	U
Formaldehyde (37%)	S	U	U
Formic acid	Q	U	U
Freons	Q	U	U
Fructose	S	Q	U
Furfural	U	U	U
Gasohol	S	Q	U
Gasolene	S	Q	U
Glucose	S	Q	U
Glycerol	S	Q	U
Heptane	S	Q	U
Hexane	S	Q	U
Hexanol	S	Q	U
Hydrobromic acid	U	U	U
Hydrochloric acid (35%)	U	U	U
Hydrochloric acid (1–25%)	U	U	U
Hydrofluoric acid	U	U	U
Hydrogen peroxide (3–10%)	Q	U	U
Hydrogen sulfide	S	U	U
Iodine	Q	U	U
Iron salts	S	Q	U
Isopropyl alcohol	S	U	U
Kerosene	S	U	U
Lead salts	S	Q	U
Linseed oil	S	U	U
Lithium salts	S	Q	U
Magnesium salts	S	Q	U
Magnesium hydroxide	S	U	U
Mercury salts	S	Q	U
Methanol	S	Q	U
Methyl ethyl ketone	U	U	U
Methyl isobutyl ketone	U	U	U
Methyl sulfate	U	U	U
Milk	S	S	U
Mineral oil	S	Q	U
Naphtha	S	Q	U
Nickel salts	S	Q	U

Table 14.1. (Continued)

Corrosive	25°C	65°C	90°C
Nitric acid (conc.)	U	U	U
Nitric acid (20%)	U	U	U
Nitric acid (10%)	U	U	U
Nitrobenzene	U	U	U
Octane	S	Q	U
Oleum	U	U	U
Oxalic acid	S	Q	U
Perchloroethylene	U	U	U
Perchloric acid	Q	U	U
Phenol	U	U	U
Phosphoric acid	Q	U	U
Phthalic acid	S	U	U
Phthalic anhydride	S	U	U
Potassium hydroxide	S	Q	U
Potassium permanganate	S	U	U
Potassium salts	S	Q	U
Pyridine	S	U	U
Silver salts	S	Q	U
Soap	S	Q	U
Sodium salts	S	Q	U
Sodium chlorate	Q	U	U
Sodium hydroxide	Q	U	U
Sodium hypochlorite	Q	U	U
Sulfur dioxide	S	U	U
Sulfur trioxide	U	U	U
Sulfuric acid (100%)	U	U	U
Sulfuric acid (70%)	U	U	U
Sulfuric acid (1–50%)	U	U	U
Tetralin	Q	U	U
Tin salts	S	Q	U
Toluene	U	U	U
Trichloroethylene	U	U	U
Turpentine	U	U	U
Urine	S	S	U
Vegetable oils	S	Q	U
Wine	S	S	U
Xylene	U	U	U
Zinc salts	S	Q	U
		Aramids	
Nonoxidizing acids (20% H_2SO_4)	U	U	U
Oxidizing acids (10% HNO_3)	U	U	U
Aqueous salt solution (NaCl)	S	S	S
Aqueous alkalies (NaOH)	S	S	Q

Table 14.1. (Continued)

Corrosive	25°C	65°C	90°C
Polar solvents (C_2H_5OH)	Q	U	U
Nonpolar solvents (C_6H_6)	S	Q	U
Water	S	S	Q
Acetaldehyde	Q	U	U
Acetic acid (glacial)	Q	U	U
Acetic acid (1–50%)	Q	U	U
Acetic anhydride	Q	U	U
Acetone	S	Q	U
Acetonitrile	S	Q	U
Acetophenone	S	U	U
Acetyl chloride	Q	U	U
Acrylonitrile	S	Q	U
Adipic acid	S	U	U
Allyl alcohol	Q	U	U
Allyl chloride	Q	U	U
Alum (aqueous solutions)	S	S	S
Aluminum salts	S	S	S
Ammonia (anhydrous)	S	U	U
Ammonia (aqueous)	S	U	U
Ammonium salts	S	S	S
Amyl acetate	S	Q	Q
Amyl alcohol	Q	U	U
Amyl chloride	Q	U	U
Aniline	U	U	U
Antimony salts	S	S	S
Aqua regia	U	U	U
Arsenic salts	S	S	S
Barium salts	S	S	S
Barium hydroxide	S	S	S
Beer	S	S	Q
Benzaldehyde	Q	U	U
Benzene	S	Q	U
Benzene sulfonic acid	U	U	U
Benzoic acid	Q	U	U
Benzyl alcohol	Q	U	U
Benzyl chloride	Q	U	U
Bismuth salts	S	S	S
Borax	S	S	S
Boric acid	S	Q	U
Bromine	U	U	U
Butadiene	S	Q	U
Butane	S	Q	U
Butyl acetate	S	Q	U

Table 14.1. **(Continued)**

Corrosive	25°C	65°C	90°C
Butyl alcohol	S	S	U
Butylamine	Q	U	U
Butyl bromide	Q	U	U
Butyl chloride	Q	U	U
Calcium salts	S	S	S
Calcium hydroxide	S	S	S
Calcium hypochlorite	Q	U	U
Carbon disulfide	Q	Q	U
Carbon tetrachloride	S	S	Q
Carbonic acid (aq. CO_2)	S	S	Q
Castor oil	S	Q	U
Caustic (aqueous)	S	S	Q
Cellosolve	S	Q	U
Chlorine	U	U	U
Chlorine dioxide	U	U	U
Chloroacetic acid	U	U	U
Chlorobenzene	Q	U	U
Chloroform	S	Q	U
Chromic acid (40%)	U	U	U
Chromic acid (10%)	U	U	U
Citric acid	Q	U	U
Clorox bleach	Q	U	U
Copper salts	S	S	S
Cresol	U	U	U
Cyclohexane	S	S	S
Cyclohexanol	S	S	Q
Cyclohexanone	S	Q	U
Detergents (general)	S	S	S
Dibutyl phthalate	Q	U	U
Diesel fuel	S	Q	S
Diethylamine	Q	U	U
Dimethylformamide	U	U	U
Dioctyl phthalate	Q	U	U
Dioxane	S	Q	Q
Esters (general)	Q	U	U
Ethyl acetate	Q	U	U
Ethanol (1–95%)	Q	U	U
Ethyl chloride	S	Q	U
Ethyl ether	S	Q	Q
Ethylene chloride	S	Q	Q
Ethylenediamine	Q	U	U
Ethylene glycol	S	S	Q
Fatty acids (general)	Q	U	U
Fluorine	U	U	U

Table 14.1. (Continued)

Corrosive	25°C	65°C	90°C
Formaldehyde (37%)	S	Q	U
Formic acid	Q	U	U
Freons	S	Q	U
Fructose	S	S	S
Furfural	Q	U	U
Gasohol	S	Q	U
Gasolene	S	Q	U
Glucose	S	S	S
Glycerol	S	S	S
Heptane	S	S	U
Hexane	S	Q	U
Hexanol	S	Q	U
Hydrobromic acid	U	U	U
Hydrochloric acid (35%)	U	U	U
Hydrochloric acid (1–25%)	U	U	U
Hydrofluoric acid	U	U	U
Hydrogen peroxide (3–10%)	U	U	U
Hydrogen sulfide	Q	U	U
Iodine	Q	U	U
Iron salts	S	S	S
Isopropyl alcohol	S	S	Q
Kerosene	S	Q	U
Lead salts	S	S	S
Linseed oil	S	Q	U
Lithium salts	S	S	S
Magnesium salts	S	S	S
Magnesium hydroxide	S	S	S
Mercury salts	S	S	S
Methanol	Q	U	U
Methyl ethyl ketone	Q	U	U
Methyl isobutyl ketone	Q	U	U
Methyl sulfate	Q	U	U
Milk	S	S	Q
Mineral oil	S	Q	U
Naphtha	S	Q	U
Nickel salts	S	S	S
Nitric acid (conc.)	U	U	U
Nitric acid (20%)	U	U	U
Nitric acid (10%)	U	U	U
Nitrobenzene	Q	U	U
Octane	S	Q	U
Oleum	U	U	U
Oxalic acid	Q	Q	Q
Perchloroethylene	Q	U	U

Table 14.1. (Continued)

Corrosive	25°C	65°C	90°C
Perchloric acid	U	U	U
Phenol	U	U	U
Phosphoric acid	U	U	U
Phthalic acid	Q	U	U
Phthalic anhydride	S	Q	U
Potassium hydroxide	S	S	Q
Potassium permanganate	Q	U	U
Potassium salts	S	S	S
Pyridine	U	U	U
Silver salts	S	S	S
Soap	S	S	S
Sodium salts	S	S	Q
Sodium chlorate	Q	U	U
Sodium hydroxide	S	S	Q
Sodium hypochlorite	Q	U	U
Sulfur dioxide	U	U	U
Sulfur trioxide	U	U	U
Sulfuric acid (100%)	U	U	U
Sulfuric acid (70%)	U	U	U
Sulfuric acid (1–50%)	U	U	U
Tetralin	S	Q	Q
Tin salts	S	S	S
Toluene	S	Q	U
Trichloroethylene	S	U	U
Turpentine	S	U	U
Urine	S	S	S
Vegetable oils	S	Q	U
Wine	S	S	S
Xylene	S	Q	U
Zinc salts	S	S	S
		Ethylcellulose	
Nonoxidizing acids (20% H_2SO_4)	Q	U	U
Oxidizing acids (10% HNO_3)	U	U	U
Aqueous salt solution (NaCl)	S	Q	U
Aqueous alkalies (NaOH)	S	U	U
Polar solvents (C_2H_5OH)	Q	U	U
Nonpolar solvents (C_6H_6)	U	U	U
Water	S	U	U
Acetaldehyde	U	U	U
Acetic acid (glacial)	U	U	U
Acetic acid (1–50%)	Q	U	U
Acetic anhydride	U	U	U

Table 14.1. (Continued)

Corrosive	25°C	65°C	90°C
Acetone	U	U	U
Acetonitrile	U	U	U
Acetophenone	U	U	U
Acetyl chloride	U	U	U
Acrylonitrile	U	U	U
Adipic acid	Q	U	U
Allyl alcohol	Q	U	U
Allyl chloride	U	U	U
Alum (aqueous solutions)	S	Q	U
Aluminum salts	S	Q	U
Ammonia (anhydrous)	Q	U	U
Ammonia (aqueous)	S	U	U
Ammonium salts	S	Q	U
Amyl acetate	U	U	U
Amyl alcohol	Q	U	U
Amyl chloride	U	U	U
Aniline	U	U	U
Antimony salts	S	Q	U
Aqua regia	U	U	U
Arsenic salts	S	Q	U
Barium salts	S	Q	U
Barium hydroxide	S	Q	U
Beer	S	Q	U
Benzaldehyde	U	U	U
Benzene	U	U	U
Benzene sulfonic acid	U	U	U
Benzoic acid	Q	U	U
Benzyl alcohol	U	U	U
Benzyl chloride	U	U	U
Bismuth salts	S	Q	U
Borax	S	Q	U
Boric acid	S	U	U
Bromine	U	U	U
Butadiene	Q	U	U
Butane	Q	U	U
Butyl acetate	U	U	U
Butyl alcohol	Q	U	U
Butylamine	U	U	U
Butyl bromide	U	U	U
Butyl chloride	U	U	U
Calcium salts	S	Q	U
Calcium hydroxide	S	Q	U
Calcium hypochlorite	Q	U	U
Carbon disulfide	U	U	U

Table 14.1. (Continued)

Corrosive	25°C	65°C	90°C
Carbon tetrachloride	U	U	U
Carbonic acid (aq. CO_2)	S	Q	U
Castor oil	S	U	U
Caustic (aqueous)	S	Q	U
Cellosolve	U	U	U
Chlorine	U	U	U
Chlorine dioxide	U	U	U
Chloroacetic acid	U	U	U
Chlorobenzene	U	U	U
Chloroform	U	U	U
Chromic acid (40%)	U	U	U
Chromic acid (10%)	U	U	U
Citric acid	Q	U	U
Clorox bleach	Q	U	U
Copper salts	S	Q	U
Cresol	U	U	U
Cyclohexane	Q	U	U
Cyclohexanol	Q	U	U
Cyclohexanone	U	U	U
Detergents (general)	S	Q	U
Dibutyl phthalate	U	U	U
Diesel fuel	S	U	U
Diethylamine	U	U	U
Dimethylformamide	U	U	U
Dioctyl phthalate	U	U	U
Dioxane	U	U	U
Esters (general)	U	U	U
Ethyl acetate	U	U	U
Ethanol (1–95%)	Q	U	U
Ethyl chloride	U	U	U
Ethyl ether	U	U	U
Ethylene chloride	U	U	U
Ethylenediamine	U	U	U
Ethylene glycol	S	Q	U
Fatty acids (general)	Q	U	U
Fluorine	U	U	U
Formaldehyde (37%)	S	U	U
Formic acid	U	U	U
Freons	Q	U	U
Fructose	S	Q	U
Furfural	U	U	U
Gasohol	S	Q	U
Gasolene	S	Q	U
Glucose	S	Q	U

Table 14.1. (Continued)

Corrosive	25°C	65°C	90°C
Glycerol	S	Q	U
Heptane	S	Q	U
Hexane	S	Q	U
Hexanol	Q	U	U
Hydrobromic acid	U	U	U
Hydrochloric acid (35%)	U	U	U
Hydrochloric acid (1–25%)	U	U	U
Hydrofluoric acid	U	U	U
Hydrogen peroxide (3–10%)	Q	U	U
Hydrogen sulfide	Q	U	U
Iodine	Q	U	U
Iron salts	S	Q	U
Isopropyl alcohol	Q	U	U
Kerosene	S	Q	U
Lead salts	S	Q	U
Linseed oil	S	Q	U
Lithium salts	S	Q	U
Magnesium salts	S	Q	U
Magnesium hydroxide	S	Q	U
Mercury salts	S	Q	U
Methanol	Q	U	U
Methyl ethyl ketone	U	U	U
Methyl isobutyl ketone	U	U	U
Methyl sulfate	U	U	U
Milk	S	Q	U
Mineral oil	S	Q	U
Naphtha	S	Q	U
Nickel salts	S	Q	U
Nitric acid (conc.)	U	U	U
Nitric acid (20%)	U	U	U
Nitric acid (10%)	U	U	U
Nitrobenzene	U	U	U
Octane	S	Q	U
Oleum	U	U	U
Oxalic acid	Q	U	U
Perchloroethylene	U	U	U
Perchloric acid	U	U	U
Phenol	U	U	U
Phosphoric acid	U	U	U
Phthalic acid	Q	U	U
Phthalic anhydride	Q	U	U
Potassium hydroxide	S	Q	U
Potassium permanganate	Q	U	U
Potassium salts	S	Q	U

Table 14.1. (Continued)

Corrosive	25°C	65°C	90°C
Pyridine	U	U	U
Silver salts	S	Q	U
Soap	S	Q	U
Sodium salts	S	Q	U
Sodium chlorate	U	U	U
Sodium hydroxide	S	Q	U
Sodium hypochlorite	U	U	U
Sulfur dioxide	U	U	U
Sulfur trioxide	U	U	U
Sulfuric acid (100%)	U	U	U
Sulfuric acid (70%)	U	U	U
Sulfuric acid (1–50%)	U	U	U
Tetralin	Q	U	U
Tin salts	S	Q	U
Toluene	U	U	U
Trichloroethylene	U	U	U
Turpentine	Q	U	U
Urine	S	Q	U
Vegetable oils	S	Q	U
Wine	S	Q	U
Xylene	U	U	U
Zinc salts	S	Q	U
		Cellulose Acetate	
Nonoxidizing acids (20% H_2SO_4)	U	U	U
Oxidizing acids (10% HNO_3)	U	U	U
Aqueous salt solution (NaCl)	S	Q	U
Aqueous alkalies (NaOH)	S	Q	U
Polar solvents (C_2H_5OH)	Q	U	U
Nonpolar solvents (C_6H_6)	U	U	U
Water	S	Q	U
Acetaldehyde	U	U	U
Acetic acid (glacial)	U	U	U
Acetic acid (1–50%)	Q	U	U
Acetic anhydride	U	U	U
Acetone	U	U	U
Acetonitrile	U	U	U
Acetophenone	U	U	U
Acetyl chloride	U	U	U
Acrylonitrile	U	U	U
Adipic acid	Q	U	U
Allyl alcohol	U	U	U
Allyl chloride	U	U	U

Table 14.1. (Continued)

Corrosive	25°C	65°C	90°C
Alum (aqueous solutions)	S	Q	U
Aluminum salts	S	Q	U
Ammonia (anhydrous)	U	U	U
Ammonia (aqueous)	U	U	U
Ammonium salts	S	Q	U
Amyl acetate	U	U	U
Amyl alcohol	U	U	U
Amyl chloride	U	U	U
Aniline	U	U	U
Antimony salts	S	Q	U
Aqua regia	U	U	U
Arsenic salts	S	Q	U
Barium salts	S	Q	U
Barium hydroxide	Q	U	U
Beer	S	Q	U
Benzaldehyde	U	U	U
Benzene	U	U	U
Benzene sulfonic acid	U	U	U
Benzoic acid	Q	U	U
Benzyl alcohol	U	U	U
Benzyl chloride	U	U	U
Bismuth salts	S	Q	U
Borax	S	Q	U
Boric acid	Q	U	U
Bromine	U	U	U
Butadiene	U	U	U
Butane	U	U	U
Butyl acetate	U	U	U
Butyl alcohol	U	U	U
Butylamine	U	U	U
Butyl bromide	U	U	U
Butyl chloride	U	U	U
Calcium salts	S	Q	U
Calcium hydroxide	Q	U	U
Calcium hypochlorite	Q	U	U
Carbon disulfide	U	U	U
Carbon tetrachloride	U	U	U
Carbonic acid (aq. CO_2)	S	Q	U
Castor oil	Q	U	U
Caustic (aqueous)	U	U	U
Cellosolve	U	U	U
Chlorine	U	U	U
Chlorine dioxide	U	U	U
Chloroacetic acid	U	U	U

Table 14.1. **(Continued)**

Corrosive	25°C	65°C	90°C
Chlorobenzene	U	U	U
Chloroform	U	U	U
Chromic acid (40%)	U	U	U
Chromic acid (10%)	U	U	U
Citric acid	Q	U	U
Clorox bleach	Q	U	U
Copper salts	S	Q	U
Cresol	U	U	U
Cyclohexane	S	U	U
Cyclohexanol	U	U	U
Cyclohexanone	U	U	U
Detergents (general)	S	Q	U
Dibutyl phthalate	U	U	U
Diesel fuel	S	Q	U
Diethylamine	U	U	U
Dimethylformamide	U	U	U
Dioctyl phthalate	U	U	U
Dioxane	U	U	U
Esters (general)	U	U	U
Ethyl acetate	U	U	U
Ethanol (1–95%)	U	U	U
Ethyl chloride	U	U	U
Ethyl ether	U	U	U
Ethylene chloride	U	U	U
Ethylenediamine	U	U	U
Ethylene glycol	S	Q	U
Fatty acids (general)	Q	U	U
Fluorine	U	U	U
Formaldehyde (37%)	S	Q	U
Formic acid	U	U	U
Freons	Q	U	U
Fructose	S	Q	U
Furfural	U	U	U
Gasohol	S	Q	U
Gasolene	S	Q	U
Glucose	S	Q	U
Glycerol	S	Q	U
Heptane	S	Q	U
Hexane	S	Q	U
Hexanol	Q	U	U
Hydrobromic acid	U	U	U
Hydrochloric acid (35%)	U	U	U
Hydrochloric acid (1–25%)	U	U	U
Hydrofluoric acid	U	U	U

Table 14.1. (Continued)

Corrosive	25°C	65°C	90°C
Hydrogen peroxide (3–10%)	Q	U	U
Hydrogen sulfide	S	U	U
Iodine	U	U	U
Iron salts	S	Q	U
Isopropyl alcohol	U	U	U
Kerosene	S	Q	U
Lead salts	S	Q	U
Linseed oil	Q	U	U
Lithium salts	S	Q	U
Magnesium salts	S	Q	U
Magnesium hydroxide	Q	U	U
Mercury salts	S	U	U
Methanol	U	U	U
Methyl ethyl ketone	U	U	U
Methyl isobutyl ketone	U	U	U
Methyl sulfate	U	U	U
Milk	S	Q	U
Mineral oil	S	Q	U
Naphtha	S	Q	U
Nickel salts	S	Q	U
Nitric acid (conc.)	U	U	U
Nitric acid (20%)	U	U	U
Nitric acid (10%)	U	U	U
Nitrobenzene	U	U	U
Octane	S	Q	U
Oleum	U	U	U
Oxalic acid	S	U	U
Perchloroethylene	U	U	U
Perchloric acid	U	U	U
Phenol	U	U	U
Phosphoric acid	U	U	U
Phthalic acid	U	U	U
Phthalic anhydride	S	U	U
Potassium hydroxide	U	U	U
Potassium permanganate	Q	U	U
Potassium salts	S	Q	U
Pyridine	U	U	U
Silver salts	S	Q	U
Soap	S	Q	U
Sodium salts	S	Q	U
Sodium chlorate	U	U	U
Sodium hydroxide	U	U	U
Sodium hypochlorite	Q	U	U
Sulfur dioxide	Q	U	U

Table 14.1. **(Continued)**

Corrosive	25°C	65°C	90°C
Sulfur trioxide	U	U	U
Sulfuric acid (100%)	U	U	U
Sulfuric acid (70%)	U	U	U
Sulfuric acid (1–50%)	U	U	U
Tetralin	S	Q	U
Tin salts	S	Q	U
Toluene	U	U	U
Trichloroethylene	U	U	U
Turpentine	U	U	U
Urine	S	Q	U
Vegetable oils	S	Q	U
Wine	S	Q	[illegible]
Xylene	S	Q	[illegible]
Zinc salts	S	Q	[illegible]
	Fluoro Plastics, Pl[illegible]		
Nonoxidizing acids (20% H_2SO_4)	S	S	[illegible]
Oxidizing acids (10% HNO_3)	S	S	[illegible]
Aqueous salt solution (NaCl)	S	S	[illegible]
Aqueous alkalies (NaOH)	S	S	[illegible]
Polar solvents (C_2H_5OH)	S	S	[illegible]
Nonpolar solvents (C_6H_6)	S	S	[illegible]
Water	S	S	[illegible]
Acetaldehyde	S	S	S
Acetic acid (glacial)	S	S	S
Acetic acid (1–50%)	S	S	S
Acetic anhydride	S	S	S
Acetone	S	S	S
Acetonitrile	S	S	S
Acetophenone	S	S	S
Acetyl chloride	S	S	S
Acrylonitrile	S	S	S
Adipic acid	S	S	S
Allyl alcohol	S	S	S
Allyl chloride	S	S	S
Alum (aqueous solutions)	S	S	S
Aluminum salts	S	S	S
Ammonia (anhydrous)	S	S	S
Ammonia (aqueous)	S	S	S
Ammonium salts	S	S	S
Amyl acetate	S	S	S
Amyl alcohol	S	S	S
Amyl chloride	S	S	S

Table 14.1. (Continued)

Corrosive	25°C	65°C	90°C
Aniline	S	S	S
Antimony salts	S	S	S
Aqua regia	S	S	Q
Arsenic salts	S	S	S
Barium salts	S	S	S
Barium hydroxide	S	S	S
Beer	S	S	S
Benzaldehyde	S	S	S
Benzene	S	S	S
Benzene sulfonic acid	S	S	S
Benzoic acid	S	S	S
Benzyl alcohol	S	S	S
Benzyl chloride	S	S	S
Bismuth salts	S	S	S
Borax	S	S	S
Boric acid	S	S	S
Bromine	S	S	S
Butadiene	S	S	S
Butane	S	S	S
Butyl acetate	S	S	S
Butyl alcohol	S	S	S
Butylamine	S	S	S
Butyl bromide	S	S	S
Butyl chloride	S	S	S
Calcium salts	S	S	S
Calcium hydroxide	S	S	S
Calcium hypochlorite	S	S	S
Carbon disulfide	S	S	S
Carbon tetrachloride	S	S	S
Carbonic acid (aq. CO_2)	S	S	S
Castor oil	S	S	S
Caustic (aqueous)	S	S	S
Cellosolve	S	S	S
Chlorine	S	S	S
Chlorine dioxide	S	S	S
Chloroacetic acid	S	S	S
Chlorobenzene	S	S	S
Chloroform	S	S	S
Chromic acid (40%)	S	S	S
Chromic acid (10%)	S	S	S
Citric acid	S	S	S
Clorox bleach	S	S	S
Copper salts	S	S	S
Cresol	S	S	S

Table 14.1. (Continued)

Corrosive	25°C	65°C	90°C
Cyclohexane	S	S	S
Cyclohexanol	S	S	S
Cyclohexanone	S	S	S
Detergents (general)	S	S	S
Dibutyl phthalate	S	S	S
Diesel fuel	S	S	S
Diethylamine	S	S	S
Dimethylformamide	S	S	S
Dioctyl phthalate	S	S	S
Dioxane	S	S	S
Esters (general)	S	S	S
Ethyl acetate	S	S	S
Ethanol (1–95%)	S	S	S
Ethyl chloride	S	S	S
Ethyl ether	S	S	S
Ethylene chloride	S	S	S
Ethylenediamine	S	S	S
Ethylene glycol	S	S	S
Fatty acids (géneral)	S	S	S
Fluorine	S	Q	Q
Formaldehyde (37%)	S	S	S
Formic acid	S	S	S
Freons	S	S	S
Fructose	S	S	S
Furfural	S	S	S
Gasohol	S	S	S
Gasolene	S	S	S
Glucose	S	S	S
Glycerol	S	S	S
Heptane	S	S	S
Hexane	S	S	S
Hexanol	S	S	S
Hydrobromic acid	S	S	S
Hydrochloric acid (35%)	S	S	S
Hydrochloric acid (1–25%)	S	S	S
Hydrofluoric acid	S	S	S
Hydrogen peroxide (3–10%)	S	S	S
Hydrogen sulfide	S	S	S
Iodine	S	S	S
Iron salts	S	S	S
Isopropyl alcohol	S	S	S
Kerosene	S	S	S
Lead salts	S	S	S
Linseed oil	S	S	S

Table 14.1. (Continued)

Corrosive	25°C	65°C	90°C
Lithium salts	S	S	S
Magnesium salts	S	S	S
Magnesium hydroxide	S	S	S
Mercury salts	S	S	S
Methanol	S	S	S
Methyl ethyl ketone	S	S	S
Methyl isobutyl ketone	S	S	S
Methyl sulfate	S	S	S
Milk	S	S	S
Mineral oil	S	S	S
Naphtha	S	S	S
Nickel salts	S	S	S
Nitric acid (conc.)	S	S	S
Nitric acid (20%)	S	S	S
Nitric acid (10%)	S	S	S
Nitrobenzene	S	S	S
Octane	S	S	S
Oleum	S	S	Q
Oxalic acid	S	S	S
Perchloroethylene	S	S	S
Perchloric acid	S	Q	Q
Phenol	S	S	S
Phosphoric acid	S	S	S
Phthalic acid	S	S	S
Phthalic anhydride	S	S	S
Potassium hydroxide	S	S	S
Potassium permanganate	S	S	Q
Potassium salts	S	S	S
Pyridine	S	S	S
Silver salts	S	S	S
Soap	S	S	S
Sodium salts	S	S	S
Sodium chlorate	S	S	S
Sodium hydroxide	S	S	S
Sodium hypochlorite	S	S	S
Sulfur dioxide	S	S	S
Sulfur trioxide	S	S	S
Sulfuric acid (100%)	S	S	S
Sulfuric acid (70%)	S	S	S
Sulfuric acid (1–50%)	S	S	S
Tetralin	S	S	S
Tin salts	S	S	S
Toluene	S	S	S
Trichloroethylene	S	S	S

Table 14.1. (Continued)

Corrosive	25°C	65°C	90°C
Turpentine	S	S	S
Urine	S	S	S
Vegetable oils	S	S	S
Wine	S	S	S
Xylene	S	S	S
Zinc salts	S	S	S
		Epoxy	
Nonoxidizing acids (20% H_2SO_4)	S	S	Q
Oxidizing acids (10% HNO_3)	U	U	U
Aqueous salt solution (NaCl)	S	S	S
Aqueous alkalies (NaOH)	S	S	Q
Polar solvents (C_2H_5OH)	S	S	S
Nonpolar solvents (C_6H_6)	S	Q	U
Water	S	S	S
Acetaldehyde	U	U	U
Acetic acid (glacial)	Q	U	U
Acetic acid (1–50%)	S	Q	U
Acetic anhydride	U	U	U
Acetone	U	U	U
Acetonitrile	U	U	U
Acetophenone	U	U	U
Acetyl chloride	Q	Q	U
Acrylonitrile	Q	Q	U
Adipic acid	S	S	S
Allyl alcohol	Q	U	U
Allyl chloride	Q	Q	U
Alum (aqueous solutions)	S	S	S
Aluminum salts	S	S	S
Ammonia (anhydrous)	S	Q	Q
Ammonia (aqueous)	S	S	S
Ammonium salts	S	S	S
Amyl acetate	Q	Q	Q
Amyl alcohol	S	U	U
Amyl chloride	Q	Q	U
Aniline	U	U	U
Antimony salts	S	S	S
Aqua regia	U	U	U
Arsenic salts	S	S	S
Barium salts	S	S	S
Barium hydroxide	S	S	S
Beer	S	S	S
Benzaldehyde	Q	U	U

Table 14.1. (Continued)

Corrosive	25°C	65°C	90°C
Benzene	Q	Q	U
Benzene sulfonic acid	S	S	Q
Benzoic acid	S	S	S
Benzyl alcohol	S	Q	U
Benzyl chloride	Q	Q	U
Bismuth salts	S	S	S
Borax	S	S	S
Boric acid	S	S	S
Bromine	U	U	U
Butadiene	S	Q	Q
Butane	S	Q	U
Butyl acetate	Q	Q	Q
Butyl alcohol	S	S	U
Butylamine	Q	Q	U
Butyl bromide	Q	Q	U
Butyl chloride	Q	Q	U
Calcium salts	S	S	S
Calcium hydroxide	S	S	S
Calcium hypochlorite	U	U	U
Carbon disulfide	Q	U	U
Carbon tetrachloride	Q	U	U
Carbonic acid (aq. CO_2)	S	S	S
Castor oil	S	Q	Q
Caustic (aqueous)	S	S	Q
Cellosolve	Q	Q	Q
Chlorine	U	U	U
Chlorine dioxide	U	U	U
Chloroacetic acid	Q	Q	U
Chlorobenzene	Q	U	U
Chloroform	Q	U	U
Chromic acid (40%)	U	U	U
Chromic acid (10%)	Q	U	U
Citric acid	S	S	S
Clorox bleach	U	U	U
Copper salts	S	S	S
Cresol	Q	Q	U
Cyclohexane	S	S	Q
Cyclohexanol	S	S	Q
Cyclohexanone	S	Q	Q
Detergents (general)	S	S	S
Dibutyl phthalate	S	S	Q
Diesel fuel	S	Q	U
Diethylamine	Q	Q	U
Dimethylformamide	U	U	U

Table 14.1. (Continued)

Corrosive	25°C	65°C	90°C
Dioctyl phthalate	S	Q	Q
Dioxane	S	Q	Q
Esters (general)	Q	Q	Q
Ethyl acetate	Q	Q	Q
Ethanol (1–95%)	S	Q	Q
Ethyl chloride	S	Q	U
Ethyl ether	Q	Q	U
Ethylene chloride	Q	Q	U
Ethylenediamine	Q	Q	U
Ethylene glycol	S	S	S
Fatty acids (general)	S	S	S
Fluorine	U	U	U
Formaldehyde (37%)	S	S	S
Formic acid	Q	U	U
Freons	S	Q	Q
Fructose	S	S	S
Furfural	U	U	U
Gasohol	S	Q	U
Gasolene	S	Q	U
Glucose	S	S	S
Glycerol	S	S	S
Heptane	S	Q	U
Hexane	S	Q	U
Hexanol	S	Q	U
Hydrobromic acid	S	S	Q
Hydrochloric acid (35%)	S	S	Q
Hydrochloric acid (1–25%)	S	S	S
Hydrofluoric acid	S	S	Q
Hydrogen peroxide (3–10%)	U	U	U
Hydrogen sulfide	S	S	Q
Iodine	Q	U	U
Iron salts	S	S	S
Isopropyl alcohol	S	U	U
Kerosene	S	Q	U
Lead salts	S	S	S
Linseed oil	S	S	Q
Lithium salts	S	S	S
Magnesium salts	S	S	S
Magnesium hydroxide	S	S	S
Mercury salts	S	S	S
Methanol	S	Q	Q
Methyl ethyl ketone	Q	U	U
Methyl isobutyl ketone	Q	U	U

Table 14.1. (Continued)

Corrosive	25°C	65°C	90°C
Methyl sulfate	Q	Q	U
Milk	S	S	S
Mineral oil	S	S	Q
Naphtha	S	Q	Q
Nickel salts	S	S	S
Nitric acid (conc.)	U	U	U
Nitric acid (20%)	U	U	U
Nitric acid (10%)	Q	U	U
Nitrobenzene	Q	U	U
Octane	S	Q	Q
Oleum	U	U	U
Oxalic acid	S	S	S
Perchloroethylene	Q	U	U
Perchloric acid	Q	U	U
Phenol	Q	U	U
Phosphoric acid	S	S	S
Phthalic acid	S	S	S
Phthalic anhydride	S	S	S
Potassium hydroxide	S	S	Q
Potassium permanganate	U	U	U
Potassium salts	S	S	S
Pyridine	U	U	U
Silver salts	S	S	S
Soap	S	S	S
Sodium salts	S	S	S
Sodium chlorate	U	U	U
Sodium hydroxide	S	S	S
Sodium hypochlorite	U	U	U
Sulfur dioxide	S	S	S
Sulfur trioxide	S	Q	U
Sulfuric acid (100%)	U	U	U
Sulfuric acid (70%)	S	Q	U
Sulfuric acid (1–50%)	S	S	Q
Tetralin	S	Q	U
Tin salts	S	S	S
Toluene	Q	Q	U
Trichloroethylene	Q	U	U
Turpentine	S	Q	U
Urine	S	S	S
Vegetable oils	S	S	Q
Wine	S	S	S
Xylene	Q	Q	U
Zinc salts	S	S	S

Table 14.1. (Continued)

Corrosive	25°C	65°C	90°C
	Fluoro Plastics PVDF, PVF, PE-CTFE		
Nonoxidizing acids (20% H_2SO_4)	S	S	S
Oxidizing acids (10% HNO_3)	S	Q	Q
Aqueous salt solution (NaCl)	S	S	S
Aqueous alkalies (NaOH)	S	S	S
Polar solvents (C_2H_5OH)	S	S	S
Nonpolar solvents (C_6H_6)	S	Q	Q
Water	S	S	S
Acetaldehyde	S	Q	Q
Acetic acid (glacial)	S	S	Q
Acetic acid (1–50%)	S	S	S
Acetic anhydride	Q	Q	Q
Acetone	S	Q	Q
Acetonitrile	S	Q	Q
Acetophenone	Q	Q	U
Acetyl chloride	S	Q	Q
Acrylonitrile	S	Q	Q
Adipic acid	S	S	S
Allyl alcohol	S	S	S
Allyl chloride	S	S	Q
Alum (aqueous solutions)	S	S	S
Aluminum salts	S	S	S
Ammonia (anhydrous)	Q	U	U
Ammonia (aqueous)	S	S	S
Ammonium salts	S	S	S
Amyl acetate	S	Q	Q
Amyl alcohol	S	S	S
Amyl chloride	S	Q	Q
Aniline	S	Q	U
Antimony salts	S	S	S
Aqua regia	S	Q	U
Arsenic salts	S	S	S
Barium salts	S	S	S
Barium hydroxide	S	S	S
Beer	S	S	S
Benzaldehyde	S	Q	Q
Benzene	S	Q	Q
Benzene sulfonic acid	S	S	Q
Benzoic acid	S	S	S
Benzyl alcohol	S	Q	Q
Benzyl chloride	S	Q	Q
Bismuth salts	S	S	S

Table 14.1. (Continued)

Corrosive	25°C	65°C	90°C
Borax	S	S	S
Boric acid	S	S	S
Bromine	S	S	S
Butadiene	S	S	S
Butane	S	S	S
Butyl acetate	S	Q	U
Butyl alcohol	S	S	S
Butylamine	Q	Q	Q
Butyl bromide	S	S	S
Butyl chloride	S	S	S
Calcium salts	S	S	S
Calcium hydroxide	S	S	S
Calcium hypochlorite	S	S	S
Carbon disulfide	S	Q	U
Carbon tetrachloride	S	S	S
Carbonic acid (aq. CO_2)	S	S	S
Castor oil	S	S	S
Caustic (aqueous)	S	S	S
Cellosolve	S	Q	Q
Chlorine	S	S	S
Chlorine dioxide	S	Q	Q
Chloroacetic acid	S	Q	Q
Chlorobenzene	S	Q	Q
Chloroform	S	S	S
Chromic acid (40%)	S	Q	U
Chromic acid (10%)	S	S	Q
Citric acid	S	S	S
Clorox bleach	S	S	S
Copper salts	S	S	S
Cresol	S	S	S
Cyclohexane	S	Q	U
Cyclohexanol	S	S	Q
Cyclohexanone	S	S	Q
Detergents (general)	S	S	S
Dibutyl phthalate	S	Q	Q
Diesel fuel	S	S	S
Diethylamine	Q	Q	U
Dimethylformamide	U	U	U
Dioctyl phthalate	S	Q	Q
Dioxane	Q	Q	U
Esters (general)	S	Q	Q
Ethyl acetate	S	Q	Q
Ethanol (1–95%)	S	S	S
Ethyl chloride	S	S	S

Table 14.1. (Continued)

Corrosive	25°C	65°C	90°C
Ethyl ether	S	Q	Q
Ethylene chloride	S	Q	Q
Ethylenediamine	Q	Q	U
Ethylene glycol	S	S	S
Fatty acids (general)	S	S	S
Fluorine	Q	Q	U
Formaldehyde (37%)	S	S	Q
Formic acid	S	Q	Q
Freons	S	Q	Q
Fructose	S	S	S
Furfural	Q	Q	U
Gasohol	S	S	S
Gasolene	S	S	S
Glucose	S	S	S
Glycerol	S	S	S
Heptane	S	S	S
Hexane	S	S	S
Hexanol	S	S	S
Hydrobromic acid	S	S	S
Hydrochloric acid (35%)	S	S	S
Hydrochloric acid (1–25%)	S	S	S
Hydrofluoric acid	S	S	Q
Hydrogen peroxide (3–10%)	S	S	S
Hydrogen sulfide	S	S	S
Iodine	S	S	S
Iron salts	S	S	S
Isopropyl alcohol	S	S	S
Kerosene	S	S	S
Lead salts	S	S	S
Linseed oil	S	S	S
Lithium salts	S	S	S
Magnesium salts	S	S	S
Magnesium hydroxide	S	S	S
Mercury salts	S	S	S
Methanol	S	S	S
Methyl ethyl ketone	Q	Q	Q
Methyl isobutyl ketone	Q	Q	Q
Methyl sulfate	S	S	Q
Milk	S	S	S
Mineral oil	S	S	S
Naphtha	S	S	S
Nickel salts	S	S	S
Nitric acid (conc.)	S	Q	U

Table 14.1. (Continued)

Corrosive	25°C	65°C	90°C
Nitric acid (20%)	S	Q	U
Nitric acid (10%)	S	Q	U
Nitrobenzene	S	Q	U
Octane	S	S	S
Oleum	U	U	U
Oxalic acid	S	S	S
Perchloroethylene	S	S	S
Perchloric acid	S	Q	U
Phenol	S	S	S
Phosphoric acid	S	S	S
Phthalic acid	S	S	S
Phthalic anhydride	S	S	S
Potassium hydroxide	S	S	S
Potassium permanganate	S	S	S
Potassium salts	S	S	S
Pyridine	S	Q	Q
Silver salts	S	S	S
Soap	S	S	S
Sodium salts	S	S	S
Sodium chlorate	S	S	Q
Sodium hydroxide	S	S	S
Sodium hypochlorite	S	S	S
Sulfur dioxide	S	S	S
Sulfur trioxide	Q	Q	U
Sulfuric acid (100%)	S	S	Q
Sulfuric acid (70%)	S	S	S
Sulfuric acid (1–50%)	S	S	S
Tetralin	S	S	S
Tin salts	S	S	S
Toluene	S	S	S
Trichloroethylene	S	S	S
Turpentine	S	S	S
Urine	S	S	S
Vegetable oils	S	S	S
Wine	S	S	S
Xylene	S	S	S
Zinc salts	S	S	S
	Furan	Carbon-Filled	Cement
Nonoxidizing acids (20% H_2SO_4)	S	S	S
Oxidizing acids (10% HNO_3)	U	U	U
Aqueous salt solution (NaCl)	S	S	S
Aqueous alkalies (NaOH)	S	S	S

Table 14.1. (Continued)

Corrosive	25°C	65°C	90°C
Polar solvents (C_2H_5OH)	S	Q	Q
Nonpolar solvents (C_6H_6)	S	S	Q
Water	S	S	S
Acetaldehyde	S	S	Q
Acetic acid (glacial)	S	Q	Q
Acetic acid (1–50%)	S	S	S
Acetic anhydride	S	Q	Q
Acetone	S	S	S
Acetonitrile	S	S	S
Acetophenone	S	S	S
Acetyl chloride	S	Q	Q
Acrylonitrile	S	S	S
Adipic acid	S	S	S
Allyl alcohol	Q	Q	U
Allyl chloride	U	U	U
Alum (aqueous solutions)	S	S	S
Aluminum salts	S	S	S
Ammonia (anhydrous)	S	S	Q
Ammonia (aqueous)	S	S	S
Ammonium salts	S	S	S
Amyl acetate	S	S	Q
Amyl alcohol	S	Q	Q
Amyl chloride	U	U	U
Aniline	Q	Q	U
Antimony salts	S	S	S
Aqua regia	U	U	U
Arsenic salts	S	S	S
Barium salts	S	S	S
Barium hydroxide	S	S	S
Beer	S	S	S
Benzaldehyde	S	S	S
Benzene	S	S	S
Benzene sulfonic acid	S	S	S
Benzoic acid	S	S	S
Benzyl alcohol	S	Q	U
Benzyl chloride	S	Q	U
Bismuth salts	S	S	S
Borax	S	S	S
Boric acid	S	S	S
Bromine	U	U	U
Butadiene	S	S	S
Butane	S	S	S
Butyl acetate	S	Q	U

Table 14.1. (Continued)

Corrosive	25°C	65°C	90°C
Butyl alcohol	Q	Q	U
Butylamine	Q	Q	U
Butyl bromide	S	Q	Q
Butyl chloride	S	Q	Q
Calcium salts	S	S	S
Calcium hydroxide	S	S	S
Calcium hypochlorite	U	U	U
Carbon disulfide	S	Q	Q
Carbon tetrachloride	S	S	Q
Carbonic acid (aq. CO_2)	S	S	S
Castor oil	S	S	Q
Caustic (aqueous)	S	S	S
Cellosolve	S	S	Q
Chlorine	Q	Q	Q
Chlorine dioxide	Q	U	U
Chloroacetic acid	S	Q	Q
Chlorobenzene	S	S	Q
Chloroform	S	S	S
Chromic acid (40%)	U	U	U
Chromic acid (10%)	U	U	U
Citric acid	S	S	S
Clorox bleach	S	Q	Q
Copper salts	S	S	S
Cresol	Q	Q	U
Cyclohexane	S	S	S
Cyclohexanol	S	S	Q
Cyclohexanone	S	Q	Q
Detergents (general)	S	S	S
Dibutyl phthalate	S	Q	Q
Diesel fuel	S	S	S
Diethylamine	S	Q	U
Dimethylformamide	Q	Q	U
Dioctyl phthalate	S	S	S
Dioxane	S	S	S
Esters (general)	S	S	S
Ethyl acetate	S	S	S
Ethanol (1–95%)	S	S	S
Ethyl chloride	S	S	S
Ethyl ether	S	S	S
Ethylene chloride	S	S	S
Ethylenediamine	S	Q	U
Ethylene glycol	S	S	S
Fatty acids (general)	S	S	S
Fluorine	U	U	U

Table 14.1. (Continued)

Corrosive	25°C	65°C	90°C
Formaldehyde (37%)	S	S	S
Formic acid	S	S	S
Freons	S	S	S
Fructose	S	S	S
Furfural	S	Q	Q
Gasohol	S	S	S
Gasolene	S	S	S
Glucose	S	S	S
Glycerol	S	S	S
Heptane	S	S	S
Hexane	S	S	S
Hexanol	S	Q	Q
Hydrobromic acid	S	S	S
Hydrochloric acid (35%)	S	S	S
Hydrochloric acid (1–25%)	S	S	S
Hydrofluoric acid	S	Q	Q
Hydrogen peroxide (3–10%)	Q	Q	U
Hydrogen sulfide	S	S	S
Iodine	S	Q	Q
Iron salts	S	S	S
Isopropyl alcohol	S	S	S
Kerosene	S	S	S
Lead salts	S	S	S
Linseed oil	S	S	S
Lithium salts	S	S	S
Magnesium salts	S	S	S
Magnesium hydroxide	S	S	S
Mercury salts	S	S	S
Methanol	S	S	S
Methyl ethyl ketone	S	Q	U
Methyl isobutyl ketone	Q	Q	U
Methyl sulfate	S	S	Q
Milk	S	S	S
Mineral oil	S	S	S
Naphtha	S	S	S
Nickel salts	S	S	S
Nitric acid (conc.)	U	U	U
Nitric acid (20%)	U	U	U
Nitric acid (10%)	Q	U	U
Nitrobenzene	S	S	S
Octane	S	S	S
Oleum	U	U	U
Oxalic acid	S	S	S
Perchloroethylene	S	S	S

Table 14.1. (Continued)

Corrosive	25°C	65°C	90°C
Perchloric acid	Q	U	U
Phenol	Q	U	U
Phosphoric acid	S	S	S
Phthalic acid	S	S	S
Phthalic anhydride	S	S	S
Potassium hydroxide	S	S	S
Potassium permanganate	S	Q	Q
Potassium salts	S	S	S
Pyridine	U	U	U
Silver salts	S	S	S
Soap	S	S	S
Sodium salts	S	S	S
Sodium chlorate	Q	Q	U
Sodium hydroxide	S	S	S
Sodium hypochlorite	Q	U	U
Sulfur dioxide	S	S	S
Sulfur trioxide	S	S	S
Sulfuric acid (100%)	U	U	U
Sulfuric acid (70%)	S	Q	U
Sulfuric acid (1–50%)	S	S	S
Tetralin	S	S	S
Tin salts	S	S	S
Toluene	S	S	S
Trichloroethylene	S	S	S
Turpentine	S	S	S
Urine	S	S	S
Vegetable oils	S	S	S
Wine	S	S	S
Xylene	S	S	S
Zinc salts	S	S	S
	Nylon,	Nylon-66,	Nylon-6
Nonoxidizing acids (20% H_2SO_4)	U	U	U
Oxidizing acids (10% HNO_3)	U	U	U
Aqueous salt solution (NaCl)	S	S	S
Aqueous alkalies (NaOH)	S	S	Q
Polar solvents (C_2H_5OH)	Q	U	U
Nonpolar solvents (C_6H_6)	S	Q	U
Water	S	S	Q
Acetaldehyde	Q	U	U
Acetic acid (glacial)	U	U	U
Acetic acid (1–50%)	Q	U	U
Acetic anhydride	Q	U	U

Table 14.1. (Continued)

Corrosive	25°C	65°C	90°C
Acetone	S	Q	U
Acetonitrile	S	Q	U
Acetophenone	S	U	U
Acetyl chloride	Q	U	U
Acrylonitrile	S	Q	U
Adipic acid	S	U	U
Allyl alcohol	Q	U	U
Allyl chloride	Q	U	U
Alum (aqueous solutions)	S	S	S
Aluminum salts	S	S	S
Ammonia (anhydrous)	S	U	U
Ammonia (aqueous)	S	U	U
Ammonium salts	S	S	S
Amyl acetate	S	Q	Q
Amyl alcohol	Q	U	U
Amyl chloride	Q	U	U
Aniline	Q	U	U
Antimony salts	U	U	U
Aqua regia	U	U	U
Arsenic salts	S	S	S
Barium salts	S	S	S
Barium hydroxide	S	S	S
Beer	S	S	S
Benzaldehyde	S	S	S
Benzene	Q	U	U
Benzene sulfonic acid	U	U	U
Benzoic acid	U	U	U
Benzyl alcohol	Q	U	U
Benzyl chloride	Q	U	U
Bismuth salts	S	S	S
Borax	S	S	S
Boric acid	S	Q	U
Bromine	U	U	U
Butadiene	S	Q	U
Butane	S	Q	U
Butyl acetate	S	Q	U
Butyl alcohol	S	S	U
Butylamine	Q	U	U
Butyl bromide	Q	U	U
Butyl chloride	Q	U	U
Calcium salts	S	S	S
Calcium hydroxide	S	S	S
Calcium hypochlorite	Q	U	U
Carbon disulfide	S	Q	U

Table 14.1. (Continued)

Corrosive	25°C	65°C	90°C
Carbon tetrachloride	S	S	Q
Carbonic acid (aq. CO_2)	S	S	Q
Castor oil	S	Q	U
Caustic (aqueous)	S	S	Q
Cellosolve	S	Q	Q
Chlorine	U	U	U
Chlorine dioxide	U	U	U
Chloroacetic acid	U	U	U
Chlorobenzene	Q	U	U
Chloroform	S	Q	U
Chromic acid (40%)	U	U	U
Chromic acid (10%)	U	U	U
Citric acid	Q	U	U
Clorox bleach	Q	U	U
Copper salts	S	S	S
Cresol	U	U	U
Cyclohexane	S	S	Q
Cyclohexanol	S	S	Q
Cyclohexanone	S	Q	U
Detergents (general)	S	S	S
Dibutyl phthalate	Q	U	U
Diesel fuel	S	S	S
Diethylamine	Q	U	U
Dimethylformamide	U	U	U
Dioctyl phthalate	Q	U	U
Dioxane	S	Q	Q
Esters (general)	S	Q	U
Ethyl acetate	S	Q	U
Ethanol (1–95%)	Q	U	U
Ethyl chloride	S	Q	U
Ethyl ether	S	Q	Q
Ethylene chloride	S	Q	U
Ethylenediamine	Q	U	U
Ethylene glycol	S	S	Q
Fatty acids (general)	Q	U	U
Fluorine	U	U	U
Formaldehyde (37%)	S	Q	U
Formic acid	U	U	U
Freons	S	Q	U
Fructose	S	S	S
Furfural	Q	U	U
Gasohol	S	Q	U
Gasolene	S	Q	U
Glucose	S	S	S

Table 14.1. (Continued)

Corrosive	25°C	65°C	90°C
Glycerol	S	S	S
Heptane	S	Q	U
Hexane	S	Q	U
Hexanol	U	U	U
Hydrobromic acid	U	U	U
Hydrochloric acid (35%)	U	U	U
Hydrochloric acid (1–25%)	U	U	U
Hydrofluoric acid	U	U	U
Hydrogen peroxide (3–10%)	U	U	U
Hydrogen sulfide	Q	U	U
Iodine	Q	U	U
Iron salts	S	S	S
Isopropyl alcohol	S	S	Q
Kerosene	S	Q	U
Lead salts	S	S	S
Linseed oil	S	Q	U
Lithium salts	S	S	S
Magnesium salts	S	S	S
Magnesium hydroxide	S	S	S
Mercury salts	S	S	S
Methanol	Q	U	U
Methyl ethyl ketone	Q	U	U
Methyl isobutyl ketone	Q	U	U
Methyl sulfate	Q	U	U
Milk	S	S	S
Mineral oil	S	S	Q
Naphtha	S	Q	U
Nickel salts	S	S	S
Nitric acid (conc.)	U	U	U
Nitric acid (20%)	U	U	U
Nitric acid (10%)	U	U	U
Nitrobenzene	S	Q	U
Octane	S	Q	U
Oleum	U	U	U
Oxalic acid	Q	Q	U
Perchloroethylene	Q	U	U
Perchloric acid	U	U	U
Phenol	U	U	U
Phosphoric acid	U	U	U
Phthalic acid	Q	U	U
Phthalic anhydride	S	Q	U
Potassium hydroxide	S	S	Q
Potassium permanganate	Q	U	U
Potassium salts	S	S	S

Table 14.1. (Continued)

Corrosive	25°C	65°C	90°C
Pyridine	U	U	U
Silver salts	S	S	S
Soap	S	S	S
Sodium salts	S	S	S
Sodium chlorate	Q	U	U
Sodium hydroxide	S	S	Q
Sodium hypochlorite	Q	U	U
Sulfur dioxide	U	U	U
Sulfur trioxide	U	U	U
Sulfuric acid (100%)	U	U	U
Sulfuric acid (70%)	U	U	U
Sulfuric acid (1–50%)	U	U	U
Tetralin	S	Q	Q
Tin salts	S	S	S
Toluene	S	Q	U
Trichloroethylene	S	U	U
Turpentine	S	U	U
Urine	S	S	S
Vegetable oils	S	Q	U
Wine	S	S	S
Xylene	S	Q	U
Zinc salts	S	S	S
		Phenolic	
Nonoxidizing acids (20% H_2SO_4)	S	S	S
Oxidizing acids (10% HNO_3)	S	Q	U
Aqueous salt solution (NaCl)	S	S	S
Aqueous alkalies (NaOH)	U	U	U
Polar solvents (C_2H_5OH)	S	S	Q
Nonpolar solvents (C_6H_6)	S	Q	U
Water	S	S	S
Acetaldehyde	S	Q	U
Acetic acid (glacial)	S	Q	U
Acetic acid (1–50%)	S	S	Q
Acetic anhydride	S	Q	U
Acetone	S	Q	Q
Acetonitrile	S	Q	U
Acetophenone	S	Q	U
Acetyl chloride	S	Q	U
Acrylonitrile	S	Q	U
Adipic acid	S	S	S
Allyl alcohol	S	Q	U
Allyl chloride	S	Q	U

Table 14.1. (Continued)

Corrosive	25°C	65°C	90°C
Alum (aqueous solutions)	S	S	S
Aluminum salts	S	S	S
Ammonia (anhydrous)	Q	U	U
Ammonia (aqueous)	Q	U	U
Ammonium salts	S	S	S
Amyl acetate	S	Q	U
Amyl alcohol	S	Q	U
Amyl chloride	S	Q	U
Aniline	Q	Q	U
Antimony salts	S	S	S
Aqua regia	U	U	U
Arsenic salts	S	S	S
Barium salts	S	S	S
Barium hydroxide	S	Q	Q
Beer	S	S	S
Benzaldehyde	S	Q	U
Benzene	S	Q	U
Benzene sulfonic acid	S	S	Q
Benzoic acid	S	S	Q
Benzyl alcohol	S	Q	U
Benzyl chloride	S	Q	U
Bismuth salts	S	S	S
Borax	S	Q	Q
Boric acid	S	S	S
Bromine	U	U	U
Butadiene	S	Q	U
Butane	S	Q	U
Butyl acetate	S	Q	U
Butyl alcohol	S	Q	U
Butylamine	S	Q	U
Butyl bromide	S	Q	U
Butyl chloride	S	Q	U
Calcium salts	S	S	S
Calcium hydroxide	Q	U	U
Calcium hypochlorite	U	U	U
Carbon disulfide	S	Q	U
Carbon tetrachloride	S	S	Q
Carbonic acid (aq. CO_2)	S	S	S
Castor oil	S	Q	Q
Caustic (aqueous)	U	U	U
Cellosolve	S	Q	U
Chlorine	U	U	U
Chlorine dioxide	U	U	U
Chloroacetic acid	S	Q	U

Table 14.1. (Continued)

Corrosive	25°C	65°C	90°C
Chlorobenzene	S	Q	U
Chloroform	S	Q	U
Chromic acid (40%)	U	U	U
Chromic acid (10%)	U	U	U
Citric acid	S	S	S
Clorox bleach	Q	U	U
Copper salts	S	S	S
Cresol	S	Q	U
Cyclohexane	S	S	U
Cyclohexanol	S	Q	U
Cyclohexanone	S	Q	U
Detergents (general)	S	Q	U
Dibutyl phthalate	S	S	U
Diesel fuel	S	Q	U
Diethylamine	S	Q	U
Dimethylformamide	S	Q	U
Dioctyl phthalate	S	Q	U
Dioxane	S	Q	U
Esters (general)	S	Q	U
Ethyl acetate	S	Q	U
Ethanol (1–95%)	S	Q	U
Ethyl chloride	S	Q	U
Ethyl ether	S	Q	U
Ethylene chloride	S	Q	U
Ethylenediamine	S	Q	U
Ethylene glycol	S	Q	U
Fatty acids (general)	S	S	S
Fluorine	S	S	Q
Formaldehyde (37%)	U	U	U
Formic acid	S	S	Q
Freons	S	S	S
Fructose	S	Q	U
Furfural	S	S	S
Gasohol	S	Q	U
Gasolene	S	S	U
Glucose	S	S	U
Glycerol	S	S	S
Heptane	S	S	S
Hexane	S	S	U
Hexanol	S	S	U
Hydrobromic acid	S	Q	U
Hydrochloric acid (35%)	S	S	S
Hydrochloric acid (1–25%)	S	S	S
Hydrofluoric acid	S	S	S

Table 14.1. (Continued)

Corrosive	25°C	65°C	90°C
Hydrogen peroxide (3–10%)	Q	Q	U
Hydrogen sulfide	S	S	Q
Iodine	Q	Q	U
Iron salts	S	S	S
Isopropyl alcohol	S	Q	U
Kerosene	S	S	U
Lead salts	S	S	S
Linseed oil	S	Q	U
Lithium salts	S	S	S
Magnesium salts	S	S	S
Magnesium hydroxide	S	Q	U
Mercury salts	S	S	S
Methanol	S	Q	U
Methyl ethyl ketone	S	Q	U
Methyl isobutyl ketone	S	Q	U
Methyl sulfate	S	Q	U
Milk	S	S	S
Mineral oil	S	S	U
Naphtha	S	S	U
Nickel salts	S	S	S
Nitric acid (conc.)	U	U	U
Nitric acid (20%)	Q	U	U
Nitric acid (10%)	S	Q	U
Nitrobenzene	S	Q	U
Octane	S	S	U
Oleum	U	U	U
Oxalic acid	S	S	Q
Perchloroethylene	S	Q	U
Perchloric acid	U	U	U
Phenol	S	Q	U
Phosphoric acid	S	S	S
Phthalic acid	S	S	Q
Phthalic anhydride	S	Q	Q
Potassium hydroxide	U	U	U
Potassium permanganate	Q	U	U
Potassium salts	S	S	S
Pyridine	Q	Q	U
Silver salts	S	S	S
Soap	S	Q	U
Sodium salts	S	S	S
Sodium chlorate	Q	U	U
Sodium hydroxide	U	U	U
Sodium hypochlorite	U	U	U
Sulfur dioxide	S	S	S
Sulfur trioxide	S	S	Q

Table 14.1. (Continued)

Corrosive	25°C	65°C	90°C
Sulfuric acid (100%)	U	U	U
Sulfuric acid (70%)	S	Q	Q
Sulfuric acid (1–50%)	S	S	S
Tetralin	S	S	U
Tin salts	S	S	S
Toluene	S	Q	U
Trichloroethylene	S	Q	U
Turpentine	S	Q	U
Urine	S	S	S
Vegetable oils	S	Q	U
Wine	S	S	S
Xylene	S	Q	U
Zinc salts	S	S	S
		Poly(Amide-Imide)	
Nonoxidizing acids (20% H_2SO_4)	S	Q	U
Oxidizing acids (10% HNO_3)	U	U	U
Aqueous salt solution (NaCl)	S	S	S
Aqueous alkalies (NaOH)	U	U	U
Polar solvents (C_2H_5OH)	S	S	S
Nonpolar solvents (C_6H_6)	S	S	S
Water	S	S	S
Acetaldehyde	S	S	S
Acetic acid (glacial)	Q	U	U
Acetic acid (1–50%)	S	Q	U
Acetic anhydride	S	S	S
Acetone	S	S	S
Acetonitrile	S	S	S
Acetophenone	S	S	S
Acetyl chloride	S	Q	U
Acrylonitrile	S	S	S
Adipic acid	S	Q	U
Allyl alcohol	S	S	Q
Allyl chloride	S	S	Q
Alum (aqueous solutions)	S	S	S
Aluminum salts	S	S	S
Ammonia (anhydrous)	U	U	U
Ammonia (aqueous)	U	U	U
Ammonium salts	S	S	S
Amyl acetate	S	S	S
Amyl alcohol	S	S	S
Amyl chloride	S	S	S
Aniline	U	U	U
Antimony salts	S	S	S

Table 14.1. **(Continued)**

Corrosive	25°C	65°C	90°C
Aqua regia	U	U	U
Arsenic salts	S	S	S
Barium salts	S	S	S
Barium hydroxide	U	U	U
Beer	S	S	S
Benzaldehyde	S	S	S
Benzene	S	S	S
Benzene sulfonic acid	S	Q	U
Benzoic acid	S	Q	U
Benzyl alcohol	S	S	S
Benzyl chloride	S	S	S
Bismuth salts	S	S	S
Borax	Q	U	U
Boric acid	S	Q	U
Bromine	S	Q	U
Butadiene	S	S	S
Butane	S	S	S
Butyl acetate	S	S	S
Butyl alcohol	S	S	S
Butylamine	Q	Q	U
Butyl bromide	S	S	S
Butyl chloride	S	S	S
Calcium salts	S	S	S
Calcium hydroxide	U	U	U
Calcium hypochlorite	U	U	U
Carbon disulfide	S	S	S
Carbon tetrachloride	S	S	S
Carbonic acid (aq. CO_2)	S	Q	Q
Castor oil	S	S	S
Caustic (aqueous)	U	U	U
Cellosolve	S	S	S
Chlorine	Q	U	U
Chlorine dioxide	U	U	U
Chloroacetic acid	Q	U	U
Chlorobenzene	S	S	S
Chloroform	S	S	S
Chromic acid (40%)	U	U	U
Chromic acid (10%)	U	U	U
Citric acid	S	Q	U
Clorox bleach	U	U	U
Copper salts	S	S	S
Cresol	S	Q	U
Cyclohexane	S	S	S
Cyclohexanol	S	S	S

Table 14.1. (Continued)

Corrosive	25°C	65°C	90°C
Cyclohexanone	S	S	S
Detergents (general)	Q	Q	U
Dibutyl phthalate	S	S	S
Diesel fuel	S	S	S
Diethylamine	Q	Q	U
Dimethylformamide	Q	U	U
Dioctyl phthalate	S	S	S
Dioxane	S	S	S
Esters (general)	S	S	S
Ethyl acetate	S	S	S
Ethanol (1–95%)	S	S	S
Ethyl chloride	S	S	S
Ethyl ether	S	S	S
Ethylene chloride	S	S	S
Ethylenediamine	Q	Q	U
Ethylene glycol	S	S	S
Fatty acids (general)	S	S	Q
Fluorine	U	U	U
Formaldehyde (37%)	S	Q	U
Formic acid	U	U	U
Freons	S	S	Q
Fructose	S	S	S
Furfural	S	Q	Q
Gasohol	S	S	S
Gasolene	S	S	S
Glucose	S	S	S
Glycerol	S	S	S
Heptane	S	S	S
Hexane	S	S	S
Hexanol	S	S	S
Hydrobromic acid	U	U	U
Hydrochloric acid (35%)	U	U	U
Hydrochloric acid (1–25%)	Q	U	U
Hydrofluoric acid	U	U	U
Hydrogen peroxide (3–10%)	S	Q	U
Hydrogen sulfide	Q	U	U
Iodine	Q	Q	U
Iron salts	S	S	S
Isopropyl alcohol	S	S	S
Kerosene	S	S	S
Lead salts	S	S	S
Linseed oil	S	S	S
Lithium salts	S	S	S
Magnesium salts	S	S	S

Table 14.1. (Continued)

Corrosive	25°C	65°C	90°C
Magnesium hydroxide	U	U	U
Mercury salts	S	S	S
Methanol	S	S	S
Methyl ethyl ketone	S	S	S
Methyl isobutyl ketone	S	S	S
Methyl sulfate	S	Q	U
Milk	S	S	S
Mineral oil	S	S	S
Naphtha	S	S	S
Nickel salts	S	S	S
Nitric acid (conc.)	U	U	U
Nitric acid (20%)	U	U	U
Nitric acid (10%)	Q	U	U
Nitrobenzene	S	U	U
Octane	S	S	S
Oleum	U	U	U
Oxalic acid	Q	U	U
Perchloroethylene	S	S	S
Perchloric acid	U	U	U
Phenol	S	Q	U
Phosphoric acid	U	U	U
Phthalic acid	Q	U	U
Phthalic anhydride	S	Q	U
Potassium hydroxide	U	U	U
Potassium permanganate	U	U	U
Potassium salts	S	S	S
Pyridine	U	U	U
Silver salts	S	S	S
Soap	S	Q	U
Sodium salts	S	S	S
Sodium chlorate	U	U	U
Sodium hydroxide	U	U	U
Sodium hypochlorite	U	U	U
Sulfur dioxide	S	Q	U
Sulfur trioxide	Q	U	U
Sulfuric acid (100%)	U	U	U
Sulfuric acid (70%)	Q	U	U
Sulfuric acid (1–50%)	Q	U	U
Tetralin	S	S	S
Tin salts	S	S	S
Toluene	S	S	S
Trichloroethylene	S	S	S
Turpentine	S	S	Q
Urine	S	S	S

Table 14.1. (Continued)

Corrosive	25°C	65°C	90°C
Vegetable oils	S	S	S
Wine	S	S	S
Xylene	S	S	S
Zinc salts	S	S	S
		Poly Aryl Esters	
Nonoxidizing acids (20% H_2SO_4)	S	S	Q
Oxidizing acids (10% HNO_3)	Q	U	U
Aqueous salt solution (NaCl)	S	S	S
Aqueous alkalies (NaOH)	S	Q	Q
Polar solvents (C_2H_5OH)	S	Q	U
Nonpolar solvents (C_6H_6)	S	S	Q
Water	S	S	S
Acetaldehyde	S	S	Q
Acetic acid (glacial)	S	Q	Q
Acetic acid (1–50%)	S	S	S
Acetic anhydride	S	S	S
Acetone	S	Q	Q
Acetonitrile	Q	U	U
Acetophenone	Q	U	U
Acetyl chloride	Q	U	U
Acrylonitrile	Q	U	U
Adipic acid	S	S	Q
Allyl alcohol	S	Q	Q
Allyl chloride	Q	Q	U
Alum (aqueous solutions)	S	S	S
Aluminum salts	S	S	S
Ammonia (anhydrous)	S	S	S
Ammonia (aqueous)	S	Q	Q
Ammonium salts	S	S	S
Amyl acetate	Q	U	U
Amyl alcohol	S	Q	Q
Amyl chloride	Q	U	U
Aniline	Q	Q	U
Antimony salts	S	S	S
Aqua regia	U	U	U
Arsenic salts	S	S	S
Barium salts	S	S	S
Barium hydroxide	S	S	Q
Beer	S	S	S
Benzaldehyde	Q	Q	U
Benzene	S	S	Q
Benzene sulfonic acid	S	Q	Q

Table 14.1. (Continued)

Corrosive	25°C	65°C	90°C
Benzoic acid	S	S	Q
Benzyl alcohol	S	S	Q
Benzyl chloride	S	S	Q
Bismuth salts	S	S	S
Borax	S	S	Q
Boric acid	S	S	S
Bromine	Q	Q	S
Butadiene	S	S	Q
Butane	S	S	S
Butyl acetate	S	Q	Q
Butyl alcohol	S	Q	Q
Butylamine	Q	U	U
Butyl bromide	Q	Q	U
Butyl chloride	Q	Q	U
Calcium salts	S	S	S
Calcium hydroxide	S	Q	Q
Calcium hypochlorite	S	Q	Q
Carbon disulfide	Q	Q	Q
Carbon tetrachloride	S	S	S
Carbonic acid (aq. CO_2)	S	S	S
Castor oil	S	Q	Q
Caustic (aqueous)	Q	Q	U
Cellosolve	S	Q	Q
Chlorine	S	Q	Q
Chlorine dioxide	S	Q	Q
Chloroacetic acid	Q	Q	U
Chlorobenzene	Q	U	U
Chloroform	S	S	S
Chromic acid (40%)	U	U	U
Chromic acid (10%)	U	U	U
Citric acid	S	S	Q
Clorox bleach	S	Q	Q
Copper salts	S	S	S
Cresol	U	U	U
Cyclohexane	S	S	Q
Cyclohexanol	S	S	Q
Cyclohexanone	S	S	Q
Detergents (general)	S	Q	Q
Dibutyl phthalate	Q	U	U
Diesel fuel	S	S	Q
Diethylamine	S	S	Q
Dimethylformamide	S	Q	Q
Dioctyl phthalate	Q	U	U
Dioxane	S	Q	Q

Table 14.1. (Continued)

Corrosive	25°C	65°C	90°C
Esters (general)	Q	U	U
Ethyl acetate	Q	U	U
Ethanol (1–95%)	S	S	Q
Ethyl chloride	Q	Q	U
Ethyl ether	S	S	Q
Ethylene chloride	S	Q	U
Ethylenediamine	S	S	S
Ethylene glycol	S	S	S
Fatty acids (general)	S	S	Q
Fluorine	U	U	U
Formaldehyde (37%)	S	Q	Q
Formic acid	S	Q	Q
Freons	S	S	Q
Fructose	S	S	S
Furfural	Q	U	U
Gasohol	S	S	S
Gasolene	S	S	Q
Glucose	S	S	S
Glycerol	S	S	S
Heptane	S	S	Q
Hexane	S	S	Q
Hexanol	S	S	Q
Hydrobromic acid	S	S	Q
Hydrochloric acid (35%)	S	S	Q
Hydrochloric acid (1–25%)	S	S	S
Hydrofluoric acid	Q	U	U
Hydrogen peroxide (3–10%)	S	S	S
Hydrogen sulfide	S	S	S
Iodine	S	Q	Q
Iron salts	S	S	S
Isopropyl alcohol	S	Q	Q
Kerosene	S	S	Q
Lead salts	S	S	S
Linseed oil	S	S	Q
Lithium salts	S	S	S
Magnesium salts	S	S	S
Magnesium hydroxide	S	Q	Q
Mercury salts	S	S	S
Methanol	S	S	Q
Methyl ethyl ketone	S	Q	U
Methyl isobutyl ketone	Q	Q	U
Methyl sulfate	Q	Q	Q
Milk	S	S	S
Mineral oil	S	Q	Q

Table 14.1. **(Continued)**

Corrosive	25°C	65°C	90°C
Naphtha	S	Q	Q
Nickel salts	S	S	S
Nitric acid (conc.)	U	U	U
Nitric acid (20%)	U	U	U
Nitric acid (10%)	Q	U	U
Nitrobenzene	S	S	Q
Octane	S	Q	Q
Oleum	U	U	U
Oxalic acid	S	S	Q
Perchloroethylene	S	S	Q
Perchloric acid	Q	U	U
Phenol	U	U	U
Phosphoric acid	S	S	Q
Phthalic acid	S	Q	Q
Phthalic anhydride	S	Q	Q
Potassium hydroxide	S	Q	Q
Potassium permanganate	S	Q	Q
Potassium salts	S	S	S
Pyridine	Q	U	U
Silver salts	S	S	S
Soap	S	S	Q
Sodium salts	S	S	S
Sodium chlorate	S	S	Q
Sodium hydroxide	S	Q	Q
Sodium hypochlorite	S	Q	Q
Sulfur dioxide	S	S	S
Sulfur trioxide	S	Q	U
Sulfuric acid (100%)	U	U	U
Sulfuric acid (70%)	Q	Q	U
Sulfuric acid (1–50%)	S	S	S
Tetralin	S	S	Q
Tin salts	S	S	S
Toluene	S	S	Q
Trichloroethylene	Q	Q	U
Turpentine	S	S	Q
Urine	S	S	S
Vegetable oils	S	S	Q
Wine	S	S	S
Xylene	S	S	Q
Zinc salts	S	S	S
		Polycarbonate	
Nonoxidizing acids (20% H_2SO_4)	Q	Q	U
Oxidizing acids (10% HNO_3)	U	U	U

Table 14.1. (Continued)

Corrosive	25°C	65°C	90°C
Aqueous salt solution (NaCl)	S	S	S
Aqueous alkalies (NaOH)	U	U	U
Polar solvents (C_2H_5OH)	S	Q	Q
Nonpolar solvents (C_6H_6)	U	U	U
Water	S	S	S
Acetaldehyde	U	U	U
Acetic acid (glacial)	Q	U	U
Acetic acid (1–50%)	Q	Q	U
Acetic anhydride	U	U	U
Acetone	U	U	U
Acetonitrile	U	U	U
Acetophenone	U	U	U
Acetyl chloride	U	U	U
Acrylonitrile	U	U	U
Adipic acid	S	S	S
Allyl alcohol	Q	U	U
Allyl chloride	U	U	U
Alum (aqueous solutions)	S	S	Q
Aluminum salts	S	S	Q
Ammonia (anhydrous)	Q	U	U
Ammonia (aqueous)	U	U	U
Ammonium salts	S	S	S
Amyl acetate	U	U	U
Amyl alcohol	Q	U	U
Amyl chloride	U	U	U
Aniline	U	U	U
Antimony salts	S	S	S
Aqua regia	U	U	U
Arsenic salts	S	S	S
Barium salts	S	S	S
Barium hydroxide	U	U	U
Beer	S	S	S
Benzaldehyde	U	U	U
Benzene	U	U	U
Benzene sulfonic acid	U	U	U
Benzoic acid	S	Q	U
Benzyl alcohol	Q	U	U
Benzyl chloride	U	U	U
Bismuth salts	S	S	S
Borax	U	U	U
Boric acid	Q	U	U
Bromine	U	U	U
Butadiene	U	U	U

Table 14.1. (Continued)

Corrosive	25°C	65°C	90°C
Butane	Q	U	U
Butyl acetate	U	U	U
Butyl alcohol	Q	U	U
Butylamine	U	U	U
Butyl bromide	U	U	U
Butyl chloride	U	U	U
Calcium salts	S	S	S
Calcium hydroxide	U	U	U
Calcium hypochlorite	U	U	U
Carbon disulfide	U	U	U
Carbon tetrachloride	U	U	U
Carbonic acid (aq. CO_2)	S	S	S
Castor oil	S	Q	Q
Caustic (aqueous)	U	U	U
Cellosolve	U	U	U
Chlorine	U	U	U
Chlorine dioxide	U	U	U
Chloroacetic acid	U	U	U
Chlorobenzene	U	U	U
Chloroform	U	U	U
Chromic acid (40%)	Q	U	U
Chromic acid (10%)	U	U	U
Citric acid	U	U	U
Clorox bleach	S	Q	U
Copper salts	S	S	S
Cresol	U	U	U
Cyclohexane	Q	U	U
Cyclohexanol	Q	U	U
Cyclohexanone	U	U	U
Detergents (general)	Q	Q	U
Dibutyl phthalate	U	U	U
Diesel fuel	Q	U	U
Diethylamine	U	U	U
Dimethylformamide	U	U	U
Dioctyl phthalate	U	U	U
Dioxane	U	U	U
Esters (general)	U	U	U
Ethyl acetate	U	U	U
Ethanol (1–95%)	S	Q	Q
Ethyl chloride	U	U	U
Ethyl ether	U	U	U
Ethylene chloride	U	U	U
Ethylenediamine	U	U	U
Ethylene glycol	S	Q	Q

Table 14.1. (Continued)

Corrosive	25°C	65°C	90°C
Fatty acids (general)	S	Q	Q
Fluorine	U	U	U
Formaldehyde (37%)	S	Q	Q
Formic acid	Q	Q	U
Freons	Q	Q	U
Fructose	S	S	S
Furfural	U	U	U
Gasohol	Q	Q	U
Gasolene	Q	Q	U
Glucose	S	S	S
Glycerol	S	S	Q
Heptane	Q	Q	U
Hexane	Q	Q	U
Hexanol	U	U	U
Hydrobromic acid	Q	U	U
Hydrochloric acid (35%)	U	U	U
Hydrochloric acid (1–25%)	Q	Q	U
Hydrofluoric acid	U	U	U
Hydrogen peroxide (3–10%)	S	S	S
Hydrogen sulfide	S	Q	Q
Iodine	Q	Q	U
Iron salts	S	S	S
Isopropyl alcohol	S	Q	Q
Kerosene	Q	Q	Q
Lead salts	S	S	S
Linseed oil	Q	Q	U
Lithium salts	S	S	S
Magnesium salts	S	S	S
Magnesium hydroxide	U	U	U
Mercury salts	S	S	S
Methanol	S	Q	Q
Methyl ethyl ketone	U	U	U
Methyl isobutyl ketone	U	U	U
Methyl sulfate	U	U	U
Milk	S	S	S
Mineral oil	Q	Q	U
Naphtha	Q	U	U
Nickel salts	S	S	S
Nitric acid (conc.)	U	U	U
Nitric acid (20%)	U	U	U
Nitric acid (10%)	U	U	U
Nitrobenzene	U	U	U
Octane	Q	Q	U
Oleum	U	U	U

Table 14.1. (Continued)

Corrosive	25°C	65°C	90°C
Oxalic acid	S	Q	Q
Perchloroethylene	U	U	U
Perchloric acid	U	U	U
Phenol	U	U	U
Phosphoric acid	S	Q	Q
Phthalic acid	S	Q	Q
Phthalic anhydride	Q	U	U
Potassium hydroxide	U	U	U
Potassium permanganate	Q	U	U
Potassium salts	S	S	S
Pyridine	U	U	U
Silver salts	S	S	S
Soap	S	Q	U
Sodium salts	S	S	S
Sodium chlorate	Q	U	U
Sodium hydroxide	U	U	U
Sodium hypochlorite	U	U	U
Sulfur dioxide	Q	U	U
Sulfur trioxide	U	U	U
Sulfuric acid (100%)	U	U	U
Sulfuric acid (70%)	Q	U	U
Sulfuric acid (1–50%)	S	Q	U
Tetralin	Q	U	U
Tin salts	S	S	S
Toluene	U	U	U
Trichloroethylene	U	U	U
Turpentine	U	U	U
Urine	S	S	S
Vegetable oils	Q	Q	Q
Wine	S	S	Q
Xylene	U	U	U
Zinc salts	S	S	S
	Polyesters (Unsaturated)		
Nonoxidizing acids (20% H_2SO_4)	Q	Q	Q
Oxidizing acids (10% HNO_3)	U	U	U
Aqueous salt solution (NaCl)	S	S	S
Aqueous alkalies (NaOH)	U	U	U
Polar solvents (C_2H_5OH)	Q	U	U
Nonpolar solvents (C_6H_6)	U	U	U
Water	S	Q	Q
Acetaldehyde	U	U	U
Acetic acid (glacial)	U	U	U

Table 14.1. (Continued)

Corrosive	25°C	65°C	90°C
Acetic acid (1–50%)	S	Q	U
Acetic anhydride	U	U	U
Acetone	U	U	U
Acetonitrile	U	U	U
Acetophenone	U	U	U
Acetyl chloride	U	U	U
Acrylonitrile	U	U	U
Adipic acid	S	S	S
Allyl alcohol	Q	U	U
Allyl chloride	U	U	U
Alum (aqueous solutions)	S	S	S
Aluminum salts	S	S	S
Ammonia (anhydrous)	Q	Q	Q
Ammonia (aqueous)	Q	Q	U
Ammonium salts	S	S	S
Amyl acetate	U	U	U
Amyl alcohol	Q	Q	U
Amyl chloride	U	U	U
Aniline	U	U	U
Antimony salts	S	S	S
Aqua regia	U	U	U
Arsenic salts	S	S	S
Barium salts	S	S	S
Barium hydroxide	U	U	U
Beer	S	S	S
Benzaldehyde	U	U	U
Benzene	U	U	U
Benzene sulfonic acid	Q	Q	U
Benzoic acid	S	Q	U
Benzyl alcohol	U	U	U
Benzyl chloride	U	U	U
Bismuth salts	S	S	S
Borax	Q	Q	Q
Boric acid	S	S	Q
Bromine	U	U	U
Butadiene	S	Q	Q
Butane	S	Q	Q
Butyl acetate	U	U	U
Butyl alcohol	Q	U	U
Butylamine	U	U	U
Butyl bromide	U	U	U
Butyl chloride	U	U	U
Calcium salts	S	S	S
Calcium hydroxide	Q	U	U

Table 14.1. (Continued)

Corrosive	25°C	65°C	90°C
Calcium hypochlorite	Q	U	U
Carbon disulfide	U	U	U
Carbon tetrachloride	Q	U	U
Carbonic acid (aq. CO_2)	S	S	S
Castor oil	S	Q	U
Caustic (aqueous)	U	U	U
Cellosolve	U	U	U
Chlorine	Q	Q	U
Chlorine dioxide	S	Q	U
Chloroacetic acid	U	U	U
Chlorobenzene	U	U	U
Chloroform	U	U	U
Chromic acid (40%)	U	U	U
Chromic acid (10%)	Q	U	U
Citric acid	S	Q	Q
Clorox bleach	Q	U	U
Copper salts	S	S	S
Cresol	U	U	U
Cyclohexane	Q	Q	U
Cyclohexanol	Q	Q	U
Cyclohexanone	U	U	U
Detergents (general)	Q	Q	U
Dibutyl phthalate	U	U	U
Diesel fuel	S	Q	U
Diethylamine	U	U	U
Dimethylformamide	U	U	U
Dioctyl phthalate	U	U	U
Dioxane	U	U	U
Esters (general)	U	U	U
Ethyl acetate	U	U	U
Ethanol (1–95%)	Q	U	U
Ethyl chloride	U	U	U
Ethyl ether	U	U	U
Ethylene chloride	U	U	U
Ethylenediamine	U	U	U
Ethylene glycol	S	S	Q
Fatty acids (general)	S	Q	Q
Fluorine	U	U	U
Formaldehyde (37%)	S	S	Q
Formic acid	Q	U	U
Freons	S	Q	Q
Fructose	S	S	S
Furfural	U	U	U
Gasohol	S	Q	U

Table 14.1. (Continued)

Corrosive	25°C	65°C	90°C
Gasolene	S	Q	U
Glucose	S	S	S
Glycerol	S	Q	U
Heptane	S	Q	U
Hexane	Q	U	U
Hexanol	S	Q	U
Hydrobromic acid	S	Q	Q
Hydrochloric acid (35%)	S	Q	Q
Hydrochloric acid (1–25%)	S	U	U
Hydrofluoric acid	S	U	U
Hydrogen peroxide (3–10%)	S	S	S
Hydrogen sulfide	S	S	Q
Iodine	Q	Q	U
Iron salts	S	S	S
Isopropyl alcohol	Q	U	U
Kerosene	S	Q	U
Lead salts	S	S	S
Linseed oil	S	Q	U
Lithium salts	S	S	S
Magnesium salts	S	S	S
Magnesium hydroxide	U	U	U
Mercury salts	S	S	S
Methanol	Q	U	U
Methyl ethyl ketone	U	U	U
Methyl isobutyl ketone	U	U	U
Methyl sulfate	Q	U	U
Milk	S	S	S
Mineral oil	S	Q	U
Naphtha	S	Q	U
Nickel salts	S	S	S
Nitric acid (conc.)	U	U	U
Nitric acid (20%)	U	U	U
Nitric acid (10%)	U	U	U
Nitrobenzene	U	U	U
Octane	S	Q	U
Oleum	U	U	U
Oxalic acid	S	S	Q
Perchloroethylene	U	U	U
Perchloric acid	U	U	U
Phenol	U	U	U
Phosphoric acid	S	S	Q
Phthalic acid	Q	Q	Q
Phthalic anhydride	U	U	U
Potassium hydroxide	U	U	U

Table 14.1. (Continued)

Corrosive	25°C	65°C	90°C
Potassium permanganate	S	Q	Q
Potassium salts	S	S	S
Pyridine	U	U	U
Silver salts	S	S	S
Soap	S	Q	U
Sodium salts	S	S	S
Sodium chlorate	Q	U	U
Sodium hydroxide	U	U	U
Sodium hypochlorite	Q	U	U
Sulfur dioxide	S	Q	Q
Sulfur trioxide	Q	Q	U
Sulfuric acid (100%)	U	U	U
Sulfuric acid (70%)	Q	U	U
Sulfuric acid (1–50%)	S	Q	Q
Tetralin	S	Q	U
Tin salts	S	S	S
Toluene	U	U	U
Trichloroethylene	U	U	U
Turpentine	U	U	U
Urine	S	S	S
Vegetable oils	S	Q	U
Wine	S	S	Q
Xylene	U	U	U
Zinc salts	S	S	S
	Polyethylenes, LPDE, LLDPE, HDPE		
Nonoxidizing acids (20% H_2SO_4)	S	S	S
Oxidizing acids (10% HNO_3)	Q	Q	U
Aqueous salt solution (NaCl)	S	S	S
Aqueous alkalies (NaOH)	S	S	S
Polar solvents (C_2H_5OH)	S	Q	Q
Nonpolar solvents (C_6H_6)	Q	Q	Q
Water	S	S	S
Acetaldehyde	Q	U	U
Acetic acid (glacial)	U	U	U
Acetic acid (1–50%)	S	S	Q
Acetic anhydride	U	U	U
Acetone	S	Q	Q
Acetonitrile	Q	Q	U
Acetophenone	Q	Q	U
Acetyl chloride	Q	Q	U
Acrylonitrile	Q	Q	U

Table 14.1. (Continued)

Corrosive	25°C	65°C	90°C
Adipic acid	S	S	Q
Allyl alcohol	Q	U	U
Allyl chloride	U	U	U
Alum (aqueous solutions)	S	S	S
Aluminum salts	S	S	S
Ammonia (anhydrous)	S	S	Q
Ammonia (aqueous)	S	S	S
Ammonium salts	S	S	S
Amyl acetate	U	U	U
Amyl alcohol	Q	U	U
Amyl chloride	U	U	U
Aniline	Q	U	U
Antimony salts	S	S	S
Aqua regia	U	U	U
Arsenic salts	S	S	S
Barium salts	S	S	S
Barium hydroxide	S	S	S
Beer	S	S	S
Benzaldehyde	S	Q	U
Benzene	Q	Q	U
Benzene sulfonic acid	S	Q	Q
Benzoic acid	S	S	S
Benzyl alcohol	U	U	U
Benzyl chloride	U	U	U
Bismuth salts	S	S	S
Borax	S	S	S
Boric acid	S	S	S
Bromine	U	U	U
Butadiene	Q	Q	U
Butane	Q	Q	U
Butyl acetate	Q	U	U
Butyl alcohol	Q	U	U
Butylamine	U	U	U
Butyl bromide	U	U	U
Butyl chloride	U	U	U
Calcium salts	S	S	S
Calcium hydroxide	S	S	S
Calcium hypochlorite	S	S	Q
Carbon disulfide	U	U	U
Carbon tetrachloride	Q	U	U
Carbonic acid (aq. CO_2)	S	S	S
Castor oil	Q	U	U
Caustic (aqueous)	S	S	S
Cellosolve	U	U	U

Table 14.1. (Continued)

Corrosive	25°C	65°C	90°C
Chlorine	U	U	U
Chlorine dioxide	Q	U	U
Chloroacetic acid	Q	Q	U
Chlorobenzene	U	U	U
Chloroform	U	U	U
Chromic acid (40%)	Q	U	U
Chromic acid (10%)	S	Q	U
Citric acid	S	S	S
Clorox bleach	S	Q	Q
Copper salts	S	S	S
Cresol	Q	U	U
Cyclohexane	Q	Q	U
Cyclohexanol	Q	U	U
Cyclohexanone	U	U	U
Detergents (general)	S	S	S
Dibutyl phthalate	U	U	U
Diesel fuel	Q	Q	U
Diethylamine	U	U	U
Dimethylformamide	S	Q	Q
Dioctyl phthalate	Q	U	U
Dioxane	Q	Q	U
Esters (general)	Q	U	U
Ethyl acetate	Q	U	U
Ethanol (1–95%)	S	Q	Q
Ethyl chloride	U	U	U
Ethyl ether	U	U	U
Ethylene chloride	U	U	U
Ethylenediamine	U	U	U
Ethylene glycol	S	S	Q
Fatty acids (general)	S	S	Q
Fluorine	Q	U	U
Formaldehyde (37%)	S	S	S
Formic acid	S	Q	Q
Freons	S	Q	Q
Fructose	S	S	S
Furfural	S	S	S
Gasohol	Q	U	U
Gasolene	Q	U	U
Glucose	S	S	S
Glycerol	S	S	S
Heptane	Q	U	U
Hexane	Q	U	U
Hexanol	Q	U	U
Hydrobromic acid	S	Q	Q

Table 14.1. (Continued)

Corrosive	25°C	65°C	90°C
Hydrochloric acid (35%)	S	S	Q
Hydrochloric acid (1–25%)	S	S	Q
Hydrofluoric acid	S	S	Q
Hydrogen peroxide (3–10%)	S	Q	Q
Hydrogen sulfide	S	S	Q
Iodine	Q	Q	U
Iron salts	S	S	S
Isopropyl alcohol	S	Q	Q
Kerosene	Q	U	U
Lead salts	S	S	S
Linseed oil	Q	U	U
Lithium salts	S	S	S
Magnesium salts	S	S	S
Magnesium hydroxide	S	S	S
Mercury salts	S	S	S
Methanol	S	S	Q
Methyl ethyl ketone	Q	Q	U
Methyl isobutyl ketone	U	U	U
Methyl sulfate	S	Q	U
Milk	S	S	S
Mineral oil	Q	Q	U
Naphtha	Q	U	U
Nickel salts	S	S	S
Nitric acid (conc.)	U	U	U
Nitric acid (20%)	Q	U	U
Nitric acid (10%)	S	Q	U
Nitrobenzene	U	U	U
Octane	Q	U	U
Oleum	U	U	U
Oxalic acid	S	S	S
Perchloroethylene	U	U	U
Perchloric acid	S	Q	U
Phenol	Q	U	U
Phosphoric acid	S	S	S
Phthalic acid	S	Q	U
Phthalic anhydride	S	Q	Q
Potassium hydroxide	S	S	S
Potassium permanganate	S	Q	Q
Potassium salts	S	S	S
Pyridine	U	U	U
Silver salts	S	S	S
Soap	S	S	S
Sodium salts	S	S	S
Sodium chlorate	S	Q	U

Table 14.1. (Continued)

Corrosive	25°C	65°C	90°C
Sodium hydroxide	S	S	S
Sodium hypochlorite	S	Q	Q
Sulfur dioxide	S	Q	Q
Sulfur trioxide	Q	U	U
Sulfuric acid (100%)	U	U	U
Sulfuric acid (70%)	S	S	Q
Sulfuric acid (1–50%)	S	S	S
Tetralin	Q	Q	U
Tin salts	S	S	S
Toluene	Q	Q	U
Trichloroethylene	U	U	U
Turpentine	Q	U	U
Urine	S	S	S
Vegetable oils	Q	Q	U
Wine	S	S	S
Xylene	Q	Q	U
Zinc salts	S	S	S
		Polyimides	
Nonoxidizing acids (20% H_2SO_4)	Q	Q	U
Oxidizing acids (10% HNO_3)	Q	U	U
Aqueous salt solution (NaCl)	S	S	S
Aqueous alkalies (NaOH)	U	U	U
Polar solvents (C_2H_5OH)	S	S	S
Nonpolar solvents (C_6H_6)	S	S	S
Water	S	S	S
Acetaldehyde	S	S	S
Acetic acid (glacial)	Q	U	U
Acetic acid (1–50%)	Q	Q	U
Acetic anhydride	S	S	S
Acetone	S	S	S
Acetonitrile	S	S	S
Acetophenone	S	S	S
Acetyl chloride	S	S	S
Acrylonitrile	S	S	S
Adipic acid	S	S	S
Allyl alcohol	S	S	S
Allyl chloride	S	S	S
Alum (aqueous solutions)	S	S	S
Aluminum salts	S	S	S
Ammonia (anhydrous)	Q	Q	U
Ammonia (aqueous)	U	U	U
Ammonium salts	S	S	S

Table 14.1. (Continued)

Corrosive	25°C	65°C	90°C
Amyl acetate	S	S	S
Amyl alcohol	S	S	S
Amyl chloride	S	S	S
Aniline	U	U	U
Antimony salts	S	S	S
Aqua regia	U	U	U
Arsenic salts	S	S	S
Barium salts	S	S	S
Barium hydroxide	Q	U	U
Beer	S	S	S
Benzaldehyde	S	S	S
Benzene	S	S	S
Benzene sulfonic acid	S	S	S
Benzoic acid	S	Q	Q
Benzyl alcohol	S	S	S
Benzyl chloride	S	S	Q
Bismuth salts	S	S	S
Borax	Q	U	U
Boric acid	S	Q	Q
Bromine	Q	Q	U
Butadiene	S	S	S
Butane	S	S	S
Butyl acetate	S	S	S
Butyl alcohol	S	S	S
Butylamine	S	Q	Q
Butyl bromide	S	S	S
Butyl chloride	S	S	S
Calcium salts	S	S	S
Calcium hydroxide	U	U	U
Calcium hypochlorite	U	U	U
Carbon disulfide	S	S	S
Carbon tetrachloride	S	S	S
Carbonic acid (aq. CO_2)	S	S	S
Castor oil	S	S	S
Caustic (aqueous)	U	U	U
Cellosolve	S	S	S
Chlorine	Q	Q	Q
Chlorine dioxide	Q	U	U
Chloroacetic acid	S	Q	Q
Chlorobenzene	S	S	S
Chloroform	S	S	S
Chromic acid (40%)	U	U	U
Chromic acid (10%)	U	U	U
Citric acid	S	S	S

Table 14.1. (Continued)

Corrosive	25°C	65°C	90°C
Clorox bleach	S	Q	Q
Copper salts	S	S	S
Cresol	S	Q	U
Cyclohexane	S	S	S
Cyclohexanol	S	S	S
Cyclohexanone	S	S	S
Detergents (general)	S	Q	Q
Dibutyl phthalate	S	S	S
Diesel fuel	S	S	S
Diethylamine	Q	Q	U
Dimethylformamide	S	Q	Q
Dioctyl phthalate	S	S	S
Dioxane	S	S	S
Esters (general)	S	S	S
Ethyl acetate	S	S	S
Ethanol (1–95%)	S	S	S
Ethyl chloride	S	S	S
Ethyl ether	S	S	S
Ethylene chloride	S	S	S
Ethylenediamine	S	Q	Q
Ethylene glycol	S	S	S
Fatty acids (general)	S	S	S
Fluorine	Q	U	U
Formaldehyde (37%)	Q	Q	U
Formic acid	Q	Q	U
Freons	S	S	S
Fructose	S	S	S
Furfural	S	S	S
Gasohol	S	S	S
Gasolene	S	S	S
Glucose	S	S	S
Glycerol	S	S	S
Heptane	S	S	S
Hexane	S	S	S
Hexanol	S	S	S
Hydrobromic acid	S	Q	U
Hydrochloric acid (35%)	Q	Q	U
Hydrochloric acid (1–25%)	S	Q	U
Hydrofluoric acid	Q	Q	U
Hydrogen peroxide (3–10%)	Q	Q	Q
Hydrogen sulfide	Q	Q	Q
Iodine	Q	Q	Q
Iron salts	S	S	S
Isopropyl alcohol	S	S	S

Table 14.1. (Continued)

Corrosive	25°C	65°C	90°C
Kerosene	S	S	S
Lead salts	S	S	S
Linseed oil	S	S	S
Lithium salts	S	S	S
Magnesium salts	S	S	S
Magnesium hydroxide	U	U	U
Mercury salts	S	S	S
Methanol	S	S	S
Methyl ethyl ketone	S	S	S
Methyl isobutyl ketone	S	S	S
Methyl sulfate	S	S	S
Milk	S	S	S
Mineral oil	S	S	S
Naphtha	S	S	S
Nickel salts	S	S	S
Nitric acid (conc.)	S	S	S
Nitric acid (20%)	U	U	U
Nitric acid (10%)	Q	U	U
Nitrobenzene	S	Q	U
Octane	S	S	S
Oleum	U	U	U
Oxalic acid	Q	Q	U
Perchloroethylene	S	S	S
Perchloric acid	U	U	U
Phenol	Q	Q	U
Phosphoric acid	U	U	U
Phthalic acid	Q	U	U
Phthalic anhydride	S	Q	Q
Potassium hydroxide	U	U	U
Potassium permanganate	U	U	U
Potassium salts	S	S	S
Pyridine	U	U	U
Silver salts	S	S	S
Soap	S	Q	U
Sodium salts	S	S	S
Sodium chlorate	U	U	U
Sodium hydroxide	U	U	U
Sodium hypochlorite	U	U	U
Sulfur dioxide	Q	Q	U
Sulfur trioxide	U	U	U
Sulfuric acid (100%)	U	U	U
Sulfuric acid (70%)	Q	U	U
Sulfuric acid (1–50%)	Q	Q	U
Tetralin	S	S	S

Table 14.1. (Continued)

Corrosive	25°C	65°C	90°C
Tin salts	S	S	S
Toluene	S	S	S
Trichloroethylene	S	S	S
Turpentine	S	S	S
Urine	S	S	S
Vegetable oils	S	S	S
Wine	S	S	S
Xylene	S	S	S
Zinc salts	S	S	S
		Polymethyl Pentene, Polypropylene	
Nonoxidizing acids (20% H_2SO_4)	S	S	S
Oxidizing acids (10% HNO_3)	Q	Q	U
Aqueous salt solution (NaCl)	S	S	S
Aqueous alkalies (NaOH)	S	S	S
Polar solvents (C_2H_5OH)	S	S	S
Nonpolar solvents (C_6H_6)	Q	Q	U
Water	S	S	S
Acetaldehyde	S	S	S
Acetic acid (glacial)	S	Q	Q
Acetic acid (1–50%)	S	S	S
Acetic anhydride	S	S	S
Acetone	S	S	S
Acetonitrile	S	Q	Q
Acetophenone	S	S	Q
Acetyl chloride	Q	U	U
Acrylonitrile	S	Q	Q
Adipic acid	S	S	S
Allyl alcohol	S	Q	U
Allyl chloride	Q	U	U
Alum (aqueous solutions)	S	S	S
Aluminum salts	S	S	S
Ammonia (anhydrous)	S	S	Q
Ammonia (aqueous)	S	S	Q
Ammonium salts	S	S	S
Amyl acetate	S	Q	U
Amyl alcohol	S	Q	Q
Amyl chloride	U	U	U
Aniline	Q	U	U
Antimony salts	S	S	S
Aqua regia	U	U	U
Arsenic salts	S	S	S

Table 14.1. (Continued)

Corrosive	25°C	65°C	90°C
Barium salts	S	S	S
Barium hydroxide	S	S	S
Beer	S	S	S
Benzaldehyde	Q	Q	U
Benzene	Q	U	U
Benzene sulfonic acid	S	Q	Q
Benzoic acid	S	Q	Q
Benzyl alcohol	U	U	U
Benzyl chloride	U	U	U
Bismuth salts	S	S	S
Borax	S	S	S
Boric acid	S	S	S
Bromine	U	U	U
Butadiene	U	U	U
Butane	U	U	U
Butyl acetate	Q	U	U
Butyl alcohol	S	Q	Q
Butylamine	S	Q	Q
Butyl bromide	U	U	U
Butyl chloride	U	U	U
Calcium salts	S	S	S
Calcium hydroxide	S	S	S
Calcium hypochlorite	S	S	Q
Carbon disulfide	U	U	U
Carbon tetrachloride	U	U	U
Carbonic acid (aq. CO_2)	S	S	S
Castor oil	S	Q	U
Caustic (aqueous)	S	S	S
Cellosolve	S	S	S
Chlorine	Q	U	U
Chlorine dioxide	Q	U	U
Chloroacetic acid	S	Q	Q
Chlorobenzene	Q	U	U
Chloroform	U	U	U
Chromic acid (40%)	U	U	U
Chromic acid (10%)	S	Q	U
Citric acid	S	S	S
Clorox bleach	S	S	S
Copper salts	S	S	S
Cresol	Q	U	U
Cyclohexane	Q	U	U
Cyclohexanol	S	Q	U
Cyclohexanone	Q	U	U
Detergents (general)	S	S	Q

Table 14.1. (Continued)

Corrosive	25°C	65°C	90°C
Dibutyl phthalate	S	Q	Q
Diesel fuel	S	Q	U
Diethylamine	U	U	U
Dimethylformamide	Q	Q	Q
Dioctyl phthalate	S	Q	Q
Dioxane	Q	Q	U
Esters (general)	Q	Q	Q
Ethyl acetate	Q	Q	Q
Ethanol (1–95%)	S	Q	Q
Ethyl chloride	Q	Q	U
Ethyl ether	U	U	U
Ethylene chloride	Q	Q	Q
Ethylenediamine	U	U	U
Ethylene glycol	S	S	S
Fatty acids (general)	S	S	S
Fluorine	Q	U	U
Formaldehyde (37%)	S	Q	Q
Formic acid	S	Q	Q
Freons	Q	U	U
Fructose	S	S	S
Furfural	Q	U	U
Gasohol	Q	Q	U
Gasolene	Q	Q	U
Glucose	S	S	S
Glycerol	S	S	S
Heptane	S	Q	U
Hexane	S	Q	U
Hexanol	S	Q	U
Hydrobromic acid	S	Q	U
Hydrochloric acid (35%)	S	Q	U
Hydrochloric acid (1–25%)	S	Q	U
Hydrofluoric acid	S	Q	U
Hydrogen peroxide (3–10%)	S	S	S
Hydrogen sulfide	S	S	Q
Iodine	Q	Q	Q
Iron salts	S	S	S
Isopropyl alcohol	S	S	Q
Kerosene	Q	Q	U
Lead salts	S	S	S
Linseed oil	S	Q	Q
Lithium salts	S	S	S
Magnesium salts	S	S	S
Magnesium hydroxide	S	S	S
Mercury salts	S	S	S

Table 14.1. **(Continued)**

Corrosive	25°C	65°C	90°C
Methanol	S	S	Q
Methyl ethyl ketone	S	Q	Q
Methyl isobutyl ketone	S	Q	Q
Methyl sulfate	S	Q	Q
Milk	S	S	S
Mineral oil	Q	Q	U
Naphtha	Q	Q	U
Nickel salts	S	S	S
Nitric acid (conc.)	Q	U	U
Nitric acid (20%)	Q	Q	U
Nitric acid (10%)	S	Q	Q
Nitrobenzene	Q	Q	U
Octane	S	Q	Q
Oleum	U	U	U
Oxalic acid	S	S	S
Perchloroethylene	U	U	U
Perchloric acid	Q	Q	U
Phenol	Q	U	U
Phosphoric acid	S	S	S
Phthalic acid	S	S	S
Phthalic anhydride	S	Q	Q
Potassium hydroxide	S	S	S
Potassium permanganate	S	S	Q
Potassium salts	S	S	S
Pyridine	U	U	U
Silver salts	S	S	S
Soap	S	S	S
Sodium salts	S	S	S
Sodium chlorate	S	Q	Q
Sodium hydroxide	S	S	S
Sodium hypochlorite	S	Q	Q
Sulfur dioxide	S	Q	U
Sulfur trioxide	Q	Q	U
Sulfuric acid (100%)	S	Q	U
Sulfuric acid (70%)	S	S	S
Sulfuric acid (1–50%)	S	S	S
Tetralin	Q	Q	U
Tin salts	S	S	S
Toluene	Q	U	U
Trichloroethylene	U	U	U
Turpentine	Q	U	U
Urine	S	S	S
Vegetable oils	Q	Q	U
Wine	S	S	S

Table 14.1. (Continued)

Corrosive	25°C	65°C	90°C
Xylene	U	U	U
Zinc salts	S	S	S
	Polyphenylene Oxide		
Nonoxidizing acids (20% H_2SO_4)	S	Q	Q
Oxidizing acids (10% HNO_3)	Q	U	U
Aqueous salt solution (NaCl)	S	S	S
Aqueous alkalies (NaOH)	S	Q	U
Polar solvents (C_2H_5OH)	S	S	Q
Nonpolar solvents (C_6H_6)	U	U	U
Water	S	S	S
Acetaldehyde	U	U	U
Acetic acid (glacial)	S	Q	U
Acetic acid (1–50%)	S	Q	Q
Acetic anhydride	Q	Q	Q
Acetone	U	U	U
Acetonitrile	U	U	U
Acetophenone	U	U	U
Acetyl chloride	U	U	U
Acrylonitrile	U	U	U
Adipic acid	S	S	S
Allyl alcohol	Q	Q	U
Allyl chloride	U	U	U
Alum (aqueous solutions)	S	S	S
Aluminum salts	S	S	S
Ammonia (anhydrous)	S	S	Q
Ammonia (aqueous)	S	Q	U
Ammonium salts	S	S	S
Amyl acetate	U	U	U
Amyl alcohol	Q	U	U
Amyl chloride	U	U	U
Aniline	U	U	U
Antimony salts	S	S	S
Aqua regia	U	U	U
Arsenic salts	S	S	S
Barium salts	S	S	S
Barium hydroxide	S	S	S
Beer	S	S	S
Benzaldehyde	U	U	U
Benzene	U	U	U
Benzene sulfonic acid	Q	U	U
Benzoic acid	S	Q	Q
Benzyl alcohol	S	Q	Q

Table 14.1. (Continued)

Corrosive	25°C	65°C	90°C
Benzyl chloride	U	U	U
Bismuth salts	S	S	S
Borax	S	S	S
Boric acid	S	S	S
Bromine	U	U	U
Butadiene	Q	U	U
Butane	Q	U	U
Butyl acetate	U	U	U
Butyl alcohol	Q	U	U
Butylamine	U	U	U
Butyl bromide	U	U	U
Butyl chloride	U	U	U
Calcium salts	S	S	S
Calcium hydroxide	S	S	S
Calcium hypochlorite	S	Q	Q
Carbon disulfide	U	U	U
Carbon tetrachloride	U	U	U
Carbonic acid (aq. CO_2)	S	S	S
Castor oil	Q	U	U
Caustic (aqueous)	S	S	S
Cellosolve	U	U	U
Chlorine	Q	Q	U
Chlorine dioxide	U	U	U
Chloroacetic acid	U	U	U
Chlorobenzene	U	U	U
Chloroform	U	U	U
Chromic acid (40%)	U	U	U
Chromic acid (10%)	Q	Q	Q
Citric acid	S	S	S
Clorox bleach	Q	Q	U
Copper salts	S	S	S
Cresol	U	U	U
Cyclohexane	U	U	U
Cyclohexanol	U	U	U
Cyclohexanone	U	U	U
Detergents (general)	S	S	S
Dibutyl phthalate	S	Q	U
Diesel fuel	U	U	U
Diethylamine	U	U	U
Dimethylformamide	U	U	U
Dioctyl phthalate	U	U	U
Dioxane	U	U	U
Esters (general)	U	U	U
Ethyl acetate	U	U	U

Table 14.1. (Continued)

Corrosive	25°C	65°C	90°C
Ethanol (1–95%)	S	S	Q
Ethyl chloride	U	U	U
Ethyl ether	U	U	U
Ethylene chloride	U	U	U
Ethylenediamine	U	U	U
Ethylene glycol	S	S	S
Fatty acids (general)	S	S	S
Fluorine	U	U	U
Formaldehyde (37%)	S	S	S
Formic acid	U	U	U
Freons	U	U	U
Fructose	S	S	S
Furfural	U	U	U
Gasohol	U	U	U
Gasolene	U	U	U
Glucose	S	S	S
Glycerol	S	S	S
Heptane	Q	Q	Q
Hexane	Q	Q	U
Hexanol	S	Q	Q
Hydrobromic acid	S	Q	U
Hydrochloric acid (35%)	S	S	Q
Hydrochloric acid (1–25%)	S	S	Q
Hydrofluoric acid	Q	Q	Q
Hydrogen peroxide (3–10%)	S	S	S
Hydrogen sulfide	S	Q	U
Iodine	U	U	U
Iron salts	S	S	S
Isopropyl alcohol	Q	Q	U
Kerosene	U	U	U
Lead salts	S	S	S
Linseed oil	S	S	Q
Lithium salts	S	S	S
Magnesium salts	S	S	S
Magnesium hydroxide	S	S	S
Mercury salts	S	S	S
Methanol	S	S	S
Methyl ethyl ketone	U	U	U
Methyl isobutyl ketone	U	U	U
Methyl sulfate	U	U	U
Milk	S	S	S
Mineral oil	S	S	U
Naphtha	U	U	U
Nickel salts	U	U	U

Table 14.1. (Continued)

Corrosive	25°C	65°C	90°C
Nitric acid (conc.)	U	U	U
Nitric acid (20%)	S	Q	U
Nitric acid (10%)	S	S	Q
Nitrobenzene	U	U	U
Octane	U	U	U
Oleum	U	U	U
Oxalic acid	U	U	U
Perchloroethylene	U	U	U
Perchloric acid	U	U	U
Phenol	U	U	U
Phosphoric acid	S	Q	Q
Phthalic acid	U	U	U
Phthalic anhydride	U	U	U
Potassium hydroxide	S	S	S
Potassium permanganate	S	Q	Q
Potassium salts	S	S	S
Pyridine	U	U	U
Silver salts	S	S	S
Soap	S	S	S
Sodium salts	S	S	S
Sodium chlorate	Q	U	U
Sodium hydroxide	S	S	S
Sodium hypochlorite	Q	U	U
Sulfur dioxide	Q	U	U
Sulfur trioxide	Q	U	U
Sulfuric acid (100%)	Q	U	U
Sulfuric acid (70%)	Q	Q	U
Sulfuric acid (1–50%)	S	Q	Q
Tetralin	U	U	U
Tin salts	S	S	S
Toluene	U	U	U
Trichloroethylene	U	U	U
Turpentine	U	U	U
Urine	S	S	S
Vegetable oils	S	S	S
Wine	S	S	S
Xylene	U	U	U
Zinc salts	S	S	S
	Polyphenylene Sulfide		
Nonoxidizing acids (20% H_2SO_4)	S	Q	U
Oxidizing acids (10% HNO_3)	U	U	U
Aqueous salt solution (NaCl)	S	S	S
Aqueous alkalies (NaOH)	S	Q	Q

Table 14.1. (Continued)

Corrosive	25°C	65°C	90°C
Polar solvents (C_2H_5OH)	S	S	S
Nonpolar solvents (C_6H_6)	S	Q	Q
Water	S	S	S
Acetaldehyde	S	Q	Q
Acetic acid (glacial)	S	Q	Q
Acetic acid (1–50%)	S	S	S
Acetic anhydride	S	S	Q
Acetone	S	S	Q
Acetonitrile	S	Q	U
Acetophenone	S	Q	U
Acetyl chloride	S	Q	Q
Acrylonitrile	S	Q	Q
Adipic acid	S	S	S
Allyl alcohol	S	S	Q
Allyl chloride	S	Q	Q
Alum (aqueous solutions)	S	S	S
Aluminum salts	S	S	S
Ammonia (anhydrous)	S	Q	Q
Ammonia (aqueous)	S	Q	Q
Ammonium salts	S	S	S
Amyl acetate	S	S	Q
Amyl alcohol	S	S	Q
Amyl chloride	S	Q	Q
Aniline	Q	Q	U
Antimony salts	S	S	S
Aqua regia	U	U	U
Arsenic salts	S	S	S
Barium salts	S	S	S
Barium hydroxide	S	S	S
Beer	S	Q	Q
Benzaldehyde	S	Q	U
Benzene	S	S	Q
Benzene sulfonic acid	S	Q	Q
Benzoic acid	S	Q	U
Benzyl alcohol	S	S	Q
Benzyl chloride	S	Q	U
Bismuth salts	S	S	S
Borax	S	S	S
Boric acid	S	Q	Q
Bromine	Q	Q	U
Butadiene	S	S	S
Butane	S	S	S
Butyl acetate	S	S	S

Table 14.1. (Continued)

Corrosive	25°C	65°C	90°C
Butyl alcohol	S	S	S
Butylamine	Q	U	U
Butyl bromide	S	S	S
Butyl chloride	S	S	S
Calcium salts	S	S	S
Calcium hydroxide	S	Q	Q
Calcium hypochlorite	Q	U	U
Carbon disulfide	S	Q	U
Carbon tetrachloride	S	Q	U
Carbonic acid (aq. CO_2)	S	S	S
Castor oil	S	Q	U
Caustic (aqueous)	S	Q	Q
Cellosolve	S	Q	Q
Chlorine	S	Q	Q
Chlorine dioxide	U	U	U
Chloroacetic acid	S	Q	Q
Chlorobenzene	S	Q	U
Chloroform	S	Q	U
Chromic acid (40%)	U	U	U
Chromic acid (10%)	U	U	U
Citric acid	S	S	S
Clorox bleach	Q	U	U
Copper salts	S	Q	Q
Cresol	U	U	U
Cyclohexane	S	Q	Q
Cyclohexanol	S	Q	Q
Cyclohexanone	S	Q	Q
Detergents (general)	S	S	S
Dibutyl phthalate	S	Q	Q
Diesel fuel	S	Q	Q
Diethylamine	U	U	U
Dimethylformamide	Q	Q	Q
Dioctyl phthalate	S	S	Q
Dioxane	S	Q	U
Esters (general)	S	Q	U
Ethyl acetate	S	Q	U
Ethanol (1–95%)	S	S	S
Ethyl chloride	S	Q	U
Ethyl ether	S	S	Q
Ethylene chloride	S	S	Q
Ethylenediamine	U	U	U
Ethylene glycol	S	S	S
Fatty acids (general)	S	S	S
Fluorine	U	U	U

Table 14.1. (Continued)

Corrosive	25°C	65°C	90°C
Formaldehyde (37%)	S	S	S
Formic acid	S	Q	Q
Freons	S	Q	Q
Fructose	S	Q	Q
Furfural	S	S	S
Gasohol	S	Q	Q
Gasolene	S	Q	Q
Glucose	S	S	S
Glycerol	S	S	S
Heptane	S	Q	Q
Hexane	S	Q	Q
Hexanol	S	Q	Q
Hydrobromic acid	S	Q	Q
Hydrochloric acid (35%)	S	Q	U
Hydrochloric acid (1–25%)	S	Q	U
Hydrofluoric acid	U	U	U
Hydrogen peroxide (3–10%)	S	S	Q
Hydrogen sulfide	U	U	U
Iodine	Q	U	U
Iron salts	S	S	S
Isopropyl alcohol	S	S	Q
Kerosene	S	Q	Q
Lead salts	S	S	S
Linseed oil	S	S	S
Lithium salts	S	S	Q
Magnesium salts	S	S	S
Magnesium hydroxide	S	Q	Q
Mercury salts	S	S	S
Methanol	S	S	S
Methyl ethyl ketone	S	Q	Q
Methyl isobutyl ketone	S	Q	Q
Methyl sulfate	S	Q	Q
Milk	S	S	S
Mineral oil	S	S	S
Naphtha	S	S	S
Nickel salts	S	S	S
Nitric acid (conc.)	U	U	U
Nitric acid (20%)	Q	U	U
Nitric acid (10%)	S	Q	U
Nitrobenzene	S	Q	Q
Octane	Q	Q	U
Oleum	U	U	U
Oxalic acid	S	S	Q
Perchloroethylene	S	Q	Q

Table 14.1. (Continued)

Corrosive	25°C	65°C	90°C
Perchloric acid	U	U	U
Phenol	S	Q	Q
Phosphoric acid	S	S	Q
Phthalic acid	S	Q	Q
Phthalic anhydride	S	Q	Q
Potassium hydroxide	S	Q	Q
Potassium permanganate	S	S	Q
Potassium salts	S	S	S
Pyridine	U	U	U
Silver salts	S	S	S
Soap	S	S	S
Sodium salts	S	S	S
Sodium chlorate	Q	U	U
Sodium hydroxide	S	Q	U
Sodium hypochlorite	S	Q	Q
Sulfur dioxide	S	Q	Q
Sulfur trioxide	S	Q	Q
Sulfuric acid (100%)	S	Q	U
Sulfuric acid (70%)	S	S	Q
Sulfuric acid (1–50%)	S	S	S
Tetralin	S	S	S
Tin salts	S	S	S
Toluene	S	S	Q
Trichloroethylene	S	Q	U
Turpentine	S	Q	Q
Urine	S	S	S
Vegetable oils	S	S	S
Wine	S	S	S
Xylene	S	S	Q
Zinc salts	S	S	S
		Polystyrene	
Nonoxidizing acids (20% H_2SO_4)	S	Q	Q
Oxidizing acids (10% HNO_3)	U	U	U
Aqueous salt solution (NaCl)	S	S	S
Aqueous alkalies (NaOH)	S	Q	U
Polar solvents (C_2H_5OH)	S	Q	U
Nonpolar solvents (C_6H_6)	U	U	U
Water	S	S	S
Acetaldehyde	U	U	U
Acetic acid (glacial)	U	U	U
Acetic acid (1–50%)	S	Q	U
Acetic anhydride	U	U	U

Table 14.1. (Continued)

Corrosive	25°C	65°C	90°C
Acetone	Q	Q	U
Acetonitrile	U	U	U
Acetophenone	U	U	U
Acetyl chloride	U	U	U
Acrylonitrile	U	U	U
Adipic acid	S	S	S
Allyl alcohol	Q	U	U
Allyl chloride	U	U	U
Alum (aqueous solutions)	S	S	S
Aluminum salts	S	S	S
Ammonia (anhydrous)	S	Q	Q
Ammonia (aqueous)	S	S	S
Ammonium salts	S	S	S
Amyl acetate	U	U	U
Amyl alcohol	Q	U	U
Amyl chloride	U	U	U
Aniline	U	U	U
Antimony salts	S	S	S
Aqua regia	U	U	U
Arsenic salts	S	S	S
Barium salts	S	S	S
Barium hydroxide	S	S	Q
Beer	S	S	S
Benzaldehyde	U	U	U
Benzene	U	U	U
Benzene sulfonic acid	Q	Q	U
Benzoic acid	S	Q	Q
Benzyl alcohol	U	U	U
Benzyl chloride	U	U	U
Bismuth salts	S	S	S
Borax	S	S	S
Boric acid	S	S	S
Bromine	U	U	U
Butadiene	Q	U	U
Butane	Q	U	U
Butyl acetate	U	U	U
Butyl alcohol	Q	Q	U
Butylamine	U	U	U
Butyl bromide	U	U	U
Butyl chloride	U	U	U
Calcium salts	S	S	S
Calcium hydroxide	S	S	Q
Calcium hypochlorite	U	U	U
Carbon disulfide	U	U	U

Table 14.1. (Continued)

Corrosive	25°C	65°C	90°C
Carbon tetrachloride	Q	U	U
Carbonic acid (aq. CO_2)	S	S	S
Castor oil	Q	U	U
Caustic (aqueous)	Q	Q	U
Cellosolve	U	U	U
Chlorine	U	U	U
Chlorine dioxide	U	U	U
Chloroacetic acid	U	U	U
Chlorobenzene	U	U	U
Chloroform	U	U	U
Chromic acid (40%)	U	U	U
Chromic acid (10%)	U	U	U
Citric acid	S	S	S
Clorox bleach	U	U	U
Copper salts	S	S	S
Cresol	U	U	U
Cyclohexane	U	U	U
Cyclohexanol	U	U	U
Cyclohexanone	U	U	U
Detergents (general)	S	S	Q
Dibutyl phthalate	U	U	U
Diesel fuel	U	U	U
Diethylamine	U	U	U
Dimethylformamide	U	U	U
Dioctyl phthalate	U	U	U
Dioxane	U	U	U
Esters (general)	U	U	U
Ethyl acetate	U	U	U
Ethanol (1–95%)	S	S	Q
Ethyl chloride	U	U	U
Ethyl ether	U	U	U
Ethylene chloride	U	U	U
Ethylenediamine	U	U	U
Ethylene glycol	S	S	S
Fatty acids (general)	S	S	S
Fluorine	U	U	U
Formaldehyde (37%)	S	Q	Q
Formic acid	U	U	U
Freons	Q	Q	Q
Fructose	S	S	S
Furfural	U	U	U
Gasohol	U	U	U
Gasolene	U	U	U
Glucose	S	S	S

Table 14.1. (Continued)

Corrosive	25°C	65°C	90°C
Glycerol	S	S	S
Heptane	U	U	U
Hexane	U	U	U
Hexanol	U	U	U
Hydrobromic acid	S	Q	Q
Hydrochloric acid (35%)	S	S	Q
Hydrochloric acid (1–25%)	S	S	S
Hydrofluoric acid	S	S	Q
Hydrogen peroxide (3–10%)	S	S	Q
Hydrogen sulfide	U	U	U
Iodine	U	U	U
Iron salts	S	S	S
Isopropyl alcohol	S	Q	Q
Kerosene	U	U	U
Lead salts	S	S	S
Linseed oil	Q	U	U
Lithium salts	S	S	S
Magnesium salts	S	S	S
Magnesium hydroxide	S	S	S
Mercury salts	S	S	S
Methanol	S	S	Q
Methyl ethyl ketone	U	U	U
Methyl isobutyl ketone	U	U	U
Methyl sulfate	U	U	U
Milk	S	S	S
Mineral oil	Q	U	U
Naphtha	U	U	U
Nickel salts	S	S	S
Nitric acid (conc.)	U	U	U
Nitric acid (20%)	U	U	U
Nitric acid (10%)	S	Q	U
Nitrobenzene	U	U	U
Octane	U	U	U
Oleum	U	U	U
Oxalic acid	S	S	Q
Perchloroethylene	U	U	U
Perchloric acid	U	U	U
Phenol	U	U	U
Phosphoric acid	S	S	Q
Phthalic acid	S	S	S
Phthalic anhydride	Q	U	U
Potassium hydroxide	S	S	Q
Potassium permanganate	Q	Q	U
Potassium salts	S	S	S

Table 14.1. (Continued)

Corrosive	25°C	65°C	90°C
Pyridine	U	U	U
Silver salts	S	S	S
Soap	S	S	S
Sodium salts	S	S	S
Sodium chlorate	Q	U	U
Sodium hydroxide	S	Q	Q
Sodium hypochlorite	S	Q	Q
Sulfur dioxide	S	Q	Q
Sulfur trioxide	Q	U	U
Sulfuric acid (100%)	U	U	U
Sulfuric acid (70%)	S	S	Q
Sulfuric acid (1–50%)	S	S	S
Tetralin	U	U	U
Tin salts	S	S	S
Toluene	U	U	U
Trichloroethylene	U	U	U
Turpentine	U	U	U
Urine	S	S	S
Vegetable oils	Q	U	U
Wine	S	S	S
Xylene	U	U	U
Zinc salts	U	U	U
		Polysulfones	
Nonoxidizing acids (20% H_2SO_4)	S	Q	U
Oxidizing acids (10% HNO_3)	U	U	U
Aqueous salt solution (NaCl)	S	S	S
Aqueous alkalies (NaOH)	S	Q	Q
Polar solvents (C_2H_5OH)	S	Q	Q
Nonpolar solvents (C_6H_6)	U	U	U
Water	S	S	S
Acetaldehyde	U	U	U
Acetic acid (glacial)	Q	Q	U
Acetic acid (1–50%)	S	S	S
Acetic anhydride	U	U	U
Acetone	S	S	S
Acetonitrile	U	U	U
Acetophenone	U	U	U
Acetyl chloride	U	U	U
Acrylonitrile	U	U	U
Adipic acid	S	S	S
Allyl alcohol	U	U	U
Allyl chloride	U	U	U

Table 14.1. (Continued)

Corrosive	25°C	65°C	90°C
Alum (aqueous solutions)	S	S	S
Aluminum salts	S	S	S
Ammonia (anhydrous)	S	S	Q
Ammonia (aqueous)	S	S	S
Ammonium salts	S	S	S
Amyl acetate	U	U	U
Amyl alcohol	U	U	U
Amyl chloride	U	U	U
Aniline	U	U	U
Antimony salts	S	S	S
Aqua regia	U	U	U
Arsenic salts	S	S	S
Barium salts	S	S	S
Barium hydroxide	S	S	S
Beer	S	S	S
Benzaldehyde	U	U	U
Benzene	U	U	U
Benzene sulfonic acid	Q	U	U
Benzoic acid	S	Q	Q
Benzyl alcohol	U	U	U
Benzyl chloride	U	U	U
Bismuth salts	S	S	S
Borax	S	S	S
Boric acid	S	Q	Q
Bromine	U	U	U
Butadiene	Q	Q	U
Butane	Q	Q	U
Butyl acetate	U	U	U
Butyl alcohol	U	U	U
Butylamine	U	U	U
Butyl bromide	U	U	U
Butyl chloride	U	U	U
Calcium salts	S	S	S
Calcium hydroxide	S	S	S
Calcium hypochlorite	S	S	S
Carbon disulfide	U	U	U
Carbon tetrachloride	Q	U	U
Carbonic acid (aq. CO_2)	S	S	S
Castor oil	S	Q	Q
Caustic (aqueous)	S	Q	Q
Cellosolve	S	S	S
Chlorine	Q	U	U
Chlorine dioxide	U	U	U
Chloroacetic acid	S	Q	U

Table 14.1. (Continued)

Corrosive	25°C	65°C	90°C
Chlorobenzene	U	U	U
Chloroform	U	U	U
Chromic acid (40%)	U	U	U
Chromic acid (10%)	U	Q	U
Citric acid	S	S	S
Clorox bleach	Q	Q	U
Copper salts	S	S	S
Cresol	U	U	U
Cyclohexane	S	S	S
Cyclohexanol	S	Q	U
Cyclohexanone	U	U	U
Detergents (general)	S	S	S
Dibutyl phthalate	U	U	U
Diesel fuel	S	Q	Q
Diethylamine	U	U	U
Dimethylformamide	U	U	U
Dioctyl phthalate	Q	Q	U
Dioxane	U	U	U
Esters (general)	U	U	U
Ethyl acetate	U	U	U
Ethanol (1–95%)	S	S	S
Ethyl chloride	U	U	U
Ethyl ether	Q	U	U
Ethylene chloride	U	U	U
Ethylenediamine	U	U	U
Ethylene glycol	S	S	S
Fatty acids (general)	S	S	S
Fluorine	U	U	U
Formaldehyde (37%)	S	Q	Q
Formic acid	Q	U	U
Freons	Q	Q	Q
Fructose	S	S	S
Furfural	U	U	U
Gasohol	Q	Q	Q
Gasolene	Q	Q	Q
Glucose	S	S	S
Glycerol	S	S	S
Heptane	S	S	Q
Hexane	S	S	Q
Hexanol	Q	U	U
Hydrobromic acid	U	U	U
Hydrochloric acid (35%)	U	U	U
Hydrochloric acid (1–25%)	S	Q	Q
Hydrofluoric acid	Q	U	U

Table 14.1. (Continued)

Corrosive	25°C	65°C	90°C
Hydrogen peroxide (3–10%)	S	S	S
Hydrogen sulfide	S	Q	Q
Iodine	Q	U	U
Iron salts	S	S	S
Isopropyl alcohol	Q	Q	Q
Kerosene	Q	Q	Q
Lead salts	S	S	S
Linseed oil	S	S	Q
Lithium salts	S	S	S
Magnesium salts	S	S	S
Magnesium hydroxide	S	S	S
Mercury salts	S	S	S
Methanol	U	U	U
Methyl ethyl ketone	U	U	U
Methyl isobutyl ketone	U	U	U
Methyl sulfate	U	U	U
Milk	S	S	S
Mineral oil	S	Q	Q
Naphtha	S	Q	Q
Nickel salts	S	S	S
Nitric acid (conc.)	U	U	U
Nitric acid (20%)	S	U	U
Nitric acid (10%)	S	Q	U
Nitrobenzene	U	U	U
Octane	S	Q	U
Oleum	U	U	U
Oxalic acid	S	S	S
Perchloroethylene	U	U	U
Perchloric acid	Q	U	U
Phenol	U	U	U
Phosphoric acid	S	S	S
Phthalic acid	S	Q	Q
Phthalic anhydride	U	U	U
Potassium hydroxide	S	S	S
Potassium permanganate	S	Q	Q
Potassium salts	S	S	S
Pyridine	U	U	U
Silver salts	S	S	S
Soap	S	S	S
Sodium salts	S	S	S
Sodium chlorate	Q	Q	U
Sodium hydroxide	S	S	Q
Sodium hypochlorite	S	S	S
Sulfur dioxide	Q	U	U

Table 14.1. (Continued)

Corrosive	25°C	65°C	90°C
Sulfur trioxide	U	U	U
Sulfuric acid (100%)	U	U	U
Sulfuric acid (70%)	S	S	S
Sulfuric acid (1–50%)	S	S	S
Tetralin	S	Q	Q
Tin salts	S	S	S
Toluene	U	U	U
Trichloroethylene	U	U	U
Turpentine	U	U	U
Urine	S	S	S
Vegetable oils	S	S	S
Wine	S	S	S
Xylene	U	U	U
Zinc salts	S	S	S
		Polyurethanes	
Nonoxidizing acids (20% H_2SO_4)	Q	U	U
Oxidizing acids (10% HNO_3)	U	U	U
Aqueous salt solution (NaCl)	S	S	S
Aqueous alkalies (NaOH)	Q	U	U
Polar solvents (C_2H_5OH)	U	U	U
Nonpolar solvents (C_6H_6)	Q	U	U
Water	S	S	S
Acetaldehyde	U	U	U
Acetic acid (glacial)	U	U	U
Acetic acid (1–50%)	U	U	U
Acetic anhydride	U	U	U
Acetone	U	U	U
Acetonitrile	U	U	U
Acetophenone	U	U	U
Acetyl chloride	U	U	U
Acrylonitrile	U	U	U
Adipic acid	U	U	U
Allyl alcohol	U	U	U
Allyl chloride	U	U	U
Alum (aqueous solutions)	S	S	S
Aluminum salts	S	S	S
Ammonia (anhydrous)	Q	U	U
Ammonia (aqueous)	U	U	U
Ammonium salts	S	S	S
Amyl acetate	U	U	U
Amyl alcohol	U	U	U
Amyl chloride	U	U	U

Table 14.1. (Continued)

Corrosive	25°C	65°C	90°C
Aniline	U	U	U
Antimony salts	U	U	U
Aqua regia	S	S	S
Arsenic salts	U	U	U
Barium salts	S	S	S
Barium hydroxide	S	S	Q
Beer	S	S	S
Benzaldehyde	U	U	U
Benzene	S	Q	U
Benzene sulfonic acid	U	U	U
Benzoic acid	S	S	S
Benzyl alcohol	U	U	U
Benzyl chloride	U	U	U
Bismuth salts	S	S	S
Borax	S	S	S
Boric acid	S	S	S
Bromine	U	U	U
Butadiene	S	S	Q
Butane	S	S	Q
Butyl acetate	U	U	U
Butyl alcohol	U	U	U
Butylamine	U	U	U
Butyl bromide	U	U	U
Butyl chloride	U	U	U
Calcium salts	S	S	S
Calcium hydroxide	Q	U	U
Calcium hypochlorite	U	U	U
Carbon disulfide	S	S	S
Carbon tetrachloride	Q	Q	Q
Carbonic acid (aq. CO_2)	S	S	S
Castor oil	S	S	S
Caustic (aqueous)	Q	U	U
Cellosolve	U	U	U
Chlorine	U	U	U
Chlorine dioxide	U	U	U
Chloroacetic acid	U	U	U
Chlorobenzene	Q	U	U
Chloroform	Q	U	U
Chromic acid (40%)	U	U	U
Chromic acid (10%)	U	U	U
Citric acid	S	S	S
Clorox bleach	U	U	U
Copper salts	S	S	S
Cresol	U	U	U

Table 14.1. (Continued)

Corrosive	25°C	65°C	90°C
Cyclohexane	S	S	Q
Cyclohexanol	S	S	Q
Cyclohexanone	U	U	U
Detergents (general)	S	S	S
Dibutyl phthalate	U	U	U
Diesel fuel	S	S	S
Diethylamine	U	U	U
Dimethylformamide	U	U	U
Dioctyl phthalate	U	U	U
Dioxane	U	U	U
Esters (general)	U	U	U
Ethyl acetate	U	U	U
Ethanol (1–95%)	Q	U	U
Ethyl chloride	U	U	U
Ethyl ether	S	S	Q
Ethylene chloride	U	U	U
Ethylenediamine	U	U	U
Ethylene glycol	S	S	S
Fatty acids (general)	S	S	S
Fluorine	U	U	U
Formaldehyde (37%)	S	S	S
Formic acid	U	U	U
Freons	S	Q	Q
Fructose	S	S	S
Furfural	U	U	U
Gasohol	S	S	S
Gasolene	S	S	S
Glucose	S	S	S
Glycerol	S	S	S
Heptane	S	S	S
Hexane	S	S	S
Hexanol	Q	Q	Q
Hydrobromic acid	U	U	U
Hydrochloric acid (35%)	U	U	U
Hydrochloric acid (1–25%)	Q	U	U
Hydrofluoric acid	U	U	U
Hydrogen peroxide (3–10%)	Q	Q	U
Hydrogen sulfide	S	S	S
Iodine	Q	U	U
Iron salts	S	S	S
Isopropyl alcohol	U	U	U
Kerosene	S	S	S
Lead salts	S	S	S
Linseed oil	S	S	S

Table 14.1. (Continued)

Corrosive	25°C	65°C	90°C
Lithium salts	S	S	S
Magnesium salts	S	S	S
Magnesium hydroxide	Q	U	U
Mercury salts	S	S	S
Methanol	Q	U	U
Methyl ethyl ketone	U	U	U
Methyl isobutyl ketone	U	U	U
Methyl sulfate	Q	U	U
Milk	S	S	S
Mineral oil	S	S	S
Naphtha	S	S	S
Nickel salts	S	S	S
Nitric acid (conc.)	U	U	U
Nitric acid (20%)	U	U	U
Nitric acid (10%)	Q	U	U
Nitrobenzene	U	U	U
Octane	S	S	S
Oleum	U	U	U
Oxalic acid	S	S	S
Perchloroethylene	U	U	U
Perchloric acid	U	U	U
Phenol	U	U	U
Phosphoric acid	Q	U	U
Phthalic acid	S	S	S
Phthalic anhydride	U	U	U
Potassium hydroxide	Q	U	U
Potassium permanganate	U	U	U
Potassium salts	S	S	S
Pyridine	U	U	U
Silver salts	S	S	S
Soap	S	S	S
Sodium salts	S	S	S
Sodium chlorate	U	U	U
Sodium hydroxide	Q	U	U
Sodium hypochlorite	U	U	U
Sulfur dioxide	S	Q	Q
Sulfur trioxide	Q	U	U
Sulfuric acid (100%)	U	U	U
Sulfuric acid (70%)	U	U	U
Sulfuric acid (1–50%)	Q	U	U
Tetralin	S	S	S
Tin salts	S	S	S
Toluene	S	S	S
Trichloroethylene	U	U	U

Table 14.1. (Continued)

Corrosive	25°C	65°C	90°C
Turpentine	S	S	S
Urine	S	S	S
Vegetable oils	S	S	S
Wine	S	S	S
Xylene	S	S	S
Zinc salts	S	S	S
		Polyvinyl Chloride (Plasticized)	
Nonoxidizing acids (20% H_2SO_4)	S	S	S
Oxidizing acids (10% HNO_3)	S	Q	Q
Aqueous salt solution (NaCl)	S	S	S
Aqueous alkalies (NaOH)	S	S	S
Polar solvents (C_2H_5OH)	S	S	S
Nonpolar solvents (C_6H_6)	U	U	U
Water	S	S	S
Acetaldehyde	U	U	U
Acetic acid (glacial)	U	U	U
Acetic acid (1–50%)	Q	U	U
Acetic anhydride	U	U	U
Acetone	U	U	U
Acetonitrile	U	U	U
Acetophenone	U	U	U
Acetyl chloride	U	U	U
Acrylonitrile	U	U	U
Adipic acid	S	S	S
Allyl alcohol	Q	U	U
Allyl chloride	U	U	U
Alum (aqueous solutions)	S	S	S
Aluminum salts	S	S	S
Ammonia (anhydrous)	S	Q	Q
Ammonia (aqueous)	S	S	S
Ammonium salts	S	S	S
Amyl acetate	U	U	U
Amyl alcohol	Q	U	U
Amyl chloride	U	U	U
Aniline	U	U	U
Antimony salts	S	S	S
Aqua regia	U	U	U
Arsenic salts	S	S	S
Barium salts	S	S	S
Barium hydroxide	S	S	S
Beer	S	S	S

Table 14.1. (Continued)

Corrosive	25°C	65°C	90°C
Benzaldehyde	U	U	U
Benzene	U	U	U
Benzene sulfonic acid	S	S	S
Benzoic acid	Q	U	U
Benzyl alcohol	Q	U	U
Benzyl chloride	U	U	U
Bismuth salts	S	S	S
Borax	S	S	S
Boric acid	S	S	S
Bromine	U	U	U
Butadiene	Q	U	U
Butane	S	S	S
Butyl acetate	U	U	U
Butyl alcohol	S	S	S
Butylamine	U	U	U
Butyl bromide	U	U	U
Butyl chloride	U	U	U
Calcium salts	S	S	S
Calcium hydroxide	S	S	S
Calcium hypochlorite	S	S	S
Carbon disulfide	U	U	U
Carbon tetrachloride	Q	U	U
Carbonic acid (aq. CO_2)	S	S	S
Castor oil	S	S	S
Caustic (aqueous)	S	S	S
Cellosolve	Q	U	U
Chlorine	S	S	Q
Chlorine dioxide	U	U	U
Chloroacetic acid	U	U	U
Chlorobenzene	U	U	U
Chloroform	U	U	U
Chromic acid (40%)	S	S	Q
Chromic acid (10%)	S	S	S
Citric acid	S	S	S
Clorox bleach	S	Q	U
Copper salts	S	S	S
Cresol	U	U	U
Cyclohexane	U	U	U
Cyclohexanol	U	U	U
Cyclohexanone	U	U	U
Detergents (general)	S	S	S
Dibutyl phthalate	U	U	U
Diesel fuel	S	Q	U
Diethylamine	U	U	U

Table 14.1. (Continued)

Corrosive	25°C	65°C	90°C
Dimethylformamide	U	U	U
Dioctyl phthalate	U	U	U
Dioxane	U	U	U
Esters (general)	U	U	U
Ethyl acetate	U	U	U
Ethanol (1–95%)	S	S	S
Ethyl chloride	U	U	U
Ethyl ether	U	U	U
Ethylene chloride	U	U	U
Ethylenediamine	U	U	U
Ethylene glycol	S	S	S
Fatty acids (general)	S	S	S
Fluorine	U	U	U
Formaldehyde (37%)	S	S	Q
Formic acid	U	U	U
Freons	U	U	U
Fructose	S	S	S
Furfural	U	U	U
Gasohol	S	Q	Q
Gasolene	S	Q	Q
Glucose	S	S	S
Glycerol	S	S	S
Heptane	Q	Q	U
Hexane	Q	Q	U
Hexanol	Q	Q	U
Hydrobromic acid	S	S	S
Hydrochloric acid (35%)	S	S	S
Hydrochloric acid (1–25%)	S	S	S
Hydrofluoric acid	S	S	Q
Hydrogen peroxide (3–10%)	S	S	S
Hydrogen sulfide	S	S	S
Iodine	Q	U	U
Iron salts	S	S	S
Isopropyl alcohol	S	S	S
Kerosene	S	Q	Q
Lead salts	S	S	S
Linseed oil	S	S	S
Lithium salts	S	S	S
Magnesium salts	S	S	S
Magnesium hydroxide	S	S	S
Mercury salts	S	S	S
Methanol	S	S	S
Methyl ethyl ketone	U	U	U
Methyl isobutyl ketone	U	U	U

Table 14.1. (Continued)

Corrosive	25°C	65°C	90°C
Methyl sulfate	U	U	U
Milk	S	S	S
Mineral oil	S	S	S
Naphtha	S	Q	Q
Nickel salts	S	S	S
Nitric acid (conc.)	S	Q	Q
Nitric acid (20%)	S	S	S
Nitric acid (10%)	S	S	S
Nitrobenzene	U	U	U
Octane	Q	Q	U
Oleum	U	U	U
Oxalic acid	S	S	S
Perchloroethylene	U	U	U
Perchloric acid	S	Q	U
Phenol	U	U	U
Phosphoric acid	S	S	S
Phthalic acid	S	S	S
Phthalic anhydride	Q	Q	Q
Potassium hydroxide	S	S	S
Potassium permanganate	S	S	Q
Potassium salts	S	S	S
Pyridine	U	U	U
Silver salts	S	S	S
Soap	S	S	S
Sodium salts	S	S	S
Sodium chlorate	S	Q	Q
Sodium hydroxide	S	S	S
Sodium hypochlorite	S	S	Q
Sulfur dioxide	S	S	S
Sulfur trioxide	S	Q	Q
Sulfuric acid (100%)	Q	U	U
Sulfuric acid (70%)	S	S	S
Sulfuric acid (1–50%)	S	S	S
Tetralin	S	Q	Q
Tin salts	S	S	S
Toluene	U	U	U
Trichloroethylene	U	U	U
Turpentine	U	U	U
Urine	S	S	S
Vegetable oils	S	S	S
Wine	S	S	S
Xylene	U	U	U
Zinc salts	S	S	S

Table 14.1. (Continued)

Corrosive	25°C	65°C	90°C
		Silicones	
Nonoxidizing acids (20% H_2SO_4)	Q	Q	U
Oxidizing acids (10% HNO_3)	U	U	U
Aqueous salt solution (NaCl)	S	S	S
Aqueous alkalies (NaOH)	S	Q	Q
Polar solvents (C_2H_5OH)	S	S	S
Nonpolar solvents (C_6H_6)	Q	U	U
Water	S	S	S
Acetaldehyde	S	S	S
Acetic acid (glacial)	S	Q	U
Acetic acid (1–50%)	S	S	S
Acetic anhydride	S	S	S
Acetone	U	U	U
Acetonitrile	S	S	S
Acetophenone	S	S	S
Acetyl chloride	S	S	S
Acrylonitrile	S	S	S
Adipic acid	S	S	S
Allyl alcohol	S	S	S
Allyl chloride	S	S	S
Alum (aqueous solutions)	S	S	S
Aluminum salts	S	S	S
Ammonia (anhydrous)	S	S	Q
Ammonia (aqueous)	S	S	S
Ammonium salts	S	S	S
Amyl acetate	Q	Q	Q
Amyl alcohol	Q	Q	Q
Amyl chloride	Q	Q	Q
Aniline	Q	U	U
Antimony salts	S	S	S
Aqua regia	U	U	U
Arsenic salts	S	S	S
Barium salts	S	S	S
Barium hydroxide	S	S	S
Beer	S	S	S
Benzaldehyde	S	Q	Q
Benzene	Q	U	U
Benzene sulfonic acid	Q	Q	U
Benzoic acid	S	Q	Q
Benzyl alcohol	Q	U	U
Benzyl chloride	Q	U	U
Bismuth salts	S	S	S

Table 14.1. (Continued)

Corrosive	25°C	65°C	90°C
Borax	S	S	S
Boric acid	S	S	S
Bromine	Q	U	U
Butadiene	Q	U	U
Butane	Q	U	U
Butyl acetate	Q	Q	U
Butyl alcohol	Q	Q	Q
Butylamine	Q	U	U
Butyl bromide	U	U	U
Butyl chloride	U	U	U
Calcium salts	S	S	S
Calcium hydroxide	S	S	S
Calcium hypochlorite	Q	U	U
Carbon disulfide	Q	Q	U
Carbon tetrachloride	Q	Q	U
Carbonic acid (aq. CO_2)	S	S	S
Castor oil	S	S	S
Caustic (aqueous)	S	S	S
Cellosolve	U	U	U
Chlorine	S	S	Q
Chlorine dioxide	U	U	U
Chloroacetic acid	S	S	Q
Chlorobenzene	Q	Q	U
Chloroform	Q	Q	U
Chromic acid (40%)	U	U	U
Chromic acid (10%)	S	Q	Q
Citric acid	S	S	S
Clorox bleach	U	U	U
Copper salts	S	S	S
Cresol	Q	Q	Q
Cyclohexane	Q	U	U
Cyclohexanol	Q	U	U
Cyclohexanone	S	S	S
Detergents (general)	U	U	U
Dibutyl phthalate	Q	U	U
Diesel fuel	Q	U	U
Diethylamine	S	Q	Q
Dimethylformamide	S	Q	Q
Dioctyl phthalate	Q	U	U
Dioxane	S	S	S
Esters (general)	Q	U	U
Ethyl acetate	Q	U	U
Ethanol (1–95%)	S	S	S
Ethyl chloride	Q	U	U

Table 14.1. (Continued)

Corrosive	25°C	65°C	90°C
Ethyl ether	Q	Q	U
Ethylene chloride	Q	U	U
Ethylenediamine	S	S	S
Ethylene glycol	S	S	S
Fatty acids (general)	S	S	Q
Fluorine	U	U	U
Formaldehyde (37%)	S	S	S
Formic acid	S	Q	Q
Freons	S	S	Q
Fructose	S	S	S
Furfural	S	S	Q
Gasohol	Q	U	U
Gasolene	Q	U	U
Glucose	S	S	S
Glycerol	S	S	S
Heptane	Q	U	U
Hexane	Q	U	U
Hexanol	Q	U	U
Hydrobromic acid	U	U	U
Hydrochloric acid (35%)	U	U	U
Hydrochloric acid (1–25%)	Q	U	U
Hydrofluoric acid	U	U	U
Hydrogen peroxide (3–10%)	S	S	S
Hydrogen sulfide	S	S	S
Iodine	Q	Q	U
Iron salts	S	S	S
Isopropyl alcohol	U	U	U
Kerosene	Q	U	U
Lead salts	S	S	S
Linseed oil	S	S	S
Lithium salts	S	S	S
Magnesium salts	S	S	S
Magnesium hydroxide	S	S	S
Mercury salts	S	S	S
Methanol	S	Q	Q
Methyl ethyl ketone	U	U	U
Methyl isobutyl ketone	U	U	U
Methyl sulfate	U	U	U
Milk	S	S	S
Mineral oil	S	S	S
Naphtha	S	Q	Q
Nickel salts	S	S	Q
Nitric acid (conc.)	U	U	U
Nitric acid (20%)	U	U	U

Table 14.1. (Continued)

Corrosive	25°C	65°C	90°C
Nitric acid (10%)	S	Q	U
Nitrobenzene	U	U	U
Octane	Q	U	U
Oleum	U	U	U
Oxalic acid	S	S	S
Perchloroethylene	U	U	U
Perchloric acid	U	U	U
Phenol	Q	Q	Q
Phosphoric acid	S	Q	Q
Phthalic acid	S	S	S
Phthalic anhydride	S	S	S
Potassium hydroxide	S	Q	Q
Potassium permanganate	S	Q	Q
Potassium salts	S	S	S
Pyridine	U	U	U
Silver salts	S	S	S
Soap	S	S	S
Sodium salts	S	S	S
Sodium chlorate	S	Q	Q
Sodium hydroxide	S	Q	Q
Sodium hypochlorite	Q	U	U
Sulfur dioxide	S	S	S
Sulfur trioxide	S	Q	Q
Sulfuric acid (100%)	U	U	U
Sulfuric acid (70%)	U	U	U
Sulfuric acid (1–50%)	Q	Q	U
Tetralin	Q	U	U
Tin salts	S	S	S
Toluene	Q	U	U
Trichloroethylene	Q	U	U
Turpentine	Q	U	U
Urine	S	S	S
Vegetable oils	S	S	S
Wine	S	S	S
Xylene	Q	U	U
Zinc salts	S	S	S
		Thermoplastic Elastomers (Styrene–Butadiene Block Copolymers)	
Nonoxidizing acids (20% H_2SO_4)	S	S	S
Oxidizing acids (10% HNO_3)	U	U	U
Aqueous salt solution (NaCl)	S	S	S
Aqueous alkalies (NaOH)	S	S	S

Table 14.1. (Continued)

Corrosive	25°C	65°C	90°C
Polar solvents (C_2H_5OH)	S	S	S
Nonpolar solvents (C_6H_6)	U	U	U
Water	S	S	S
Acetaldehyde	U	U	U
Acetic acid (glacial)	S	Q	U
Acetic acid (1–50%)	S	Q	Q
Acetic anhydride	U	U	U
Acetone	Q	U	U
Acetonitrile	U	U	U
Acetophenone	U	U	U
Acetyl chloride	U	U	U
Acrylonitrile	U	U	U
Adipic acid	S	S	S
Allyl alcohol	U	U	U
Allyl chloride	U	U	U
Alum (aqueous solutions)	S	S	S
Aluminum salts	S	S	S
Ammonia (anhydrous)	S	Q	Q
Ammonia (aqueous)	S	S	S
Ammonium salts	S	S	S
Amyl acetate	U	U	U
Amyl alcohol	U	U	U
Amyl chloride	U	U	U
Aniline	Q	U	U
Antimony salts	S	S	S
Aqua regia	U	U	U
Arsenic salts	S	S	S
Barium salts	S	S	S
Barium hydroxide	S	S	S
Beer	S	S	S
Benzaldehyde	U	U	U
Benzene	U	U	U
Benzene sulfonic acid	S	Q	Q
Benzoic acid	S	S	S
Benzyl alcohol	Q	Q	U
Benzyl chloride	U	U	U
Bismuth salts	S	S	S
Borax	S	S	S
Boric acid	S	S	S
Bromine	U	U	U
Butadiene	U	U	U
Butane	U	U	U
Butyl acetate	U	U	U

Table 14.1. **(Continued)**

Corrosive	25°C	65°C	90°C
Butyl alcohol	Q	U	U
Butylamine	U	U	U
Butyl bromide	U	U	U
Butyl chloride	U	U	U
Calcium salts	S	S	S
Calcium hydroxide	S	S	S
Calcium hypochlorite	Q	U	U
Carbon disulfide	U	U	U
Carbon tetrachloride	U	U	U
Carbonic acid (aq. CO_2)	S	S	S
Castor oil	U	U	U
Caustic (aqueous)	S	S	S
Cellosolve	U	U	U
Chlorine	U	U	U
Chlorine dioxide	U	U	U
Chloroacetic acid	U	U	U
Chlorobenzene	U	U	U
Chloroform	U	U	U
Chromic acid (40%)	U	U	U
Chromic acid (10%)	Q	U	U
Citric acid	S	S	S
Clorox bleach	Q	U	U
Copper salts	S	S	S
Cresol	U	U	U
Cyclohexane	U	U	U
Cyclohexanol	U	U	U
Cyclohexanone	S	S	S
Detergents (general)	U	U	U
Dibutyl phthalate	U	U	U
Diesel fuel	U	U	U
Diethylamine	U	U	U
Dimethylformamide	U	U	U
Dioctyl phthalate	U	U	U
Dioxane	U	U	U
Esters (general)	U	U	U
Ethyl acetate	U	U	U
Ethanol (1–95%)	S	S	S
Ethyl chloride	U	U	U
Ethyl ether	U	U	U
Ethylene chloride	U	U	U
Ethylenediamine	U	U	U
Ethylene glycol	S	S	S
Fatty acids (general)	S	Q	Q
Fluorine	U	U	U

Table 14.1. (Continued)

Corrosive	25°C	65°C	90°C
Formaldehyde (37%)	S	S	S
Formic acid	Q	U	U
Freons	U	U	U
Fructose	S	S	S
Furfural	U	U	U
Gasohol	U	U	U
Gasolene	U	U	U
Glucose	S	S	S
Glycerol	S	S	S
Heptane	U	U	U
Hexane	U	U	U
Hexanol	U	U	U
Hydrobromic acid	U	U	U
Hydrochloric acid (35%)	Q	U	U
Hydrochloric acid (1–25%)	S	Q	Q
Hydrofluoric acid	Q	U	U
Hydrogen peroxide (3–10%)	Q	Q	Q
Hydrogen sulfide	S	S	S
Iodine	Q	U	U
Iron salts	S	S	S
Isopropyl alcohol	S	S	S
Kerosene	U	U	U
Lead salts	S	S	S
Linseed oil	U	U	U
Lithium salts	S	S	S
Magnesium salts	S	S	S
Magnesium hydroxide	S	S	S
Mercury salts	S	S	S
Methanol	S	S	S
Methyl ethyl ketone	U	U	U
Methyl isobutyl ketone	U	U	U
Methyl sulfate	U	U	U
Milk	S	S	S
Mineral oil	U	U	U
Naphtha	U	U	U
Nickel salts	S	S	S
Nitric acid (conc.)	U	U	U
Nitric acid (20%)	U	U	U
Nitric acid (10%)	Q	U	U
Nitrobenzene	U	U	U
Octane	U	U	U
Oleum	U	U	U
Oxalic acid	S	S	S
Perchloroethylene	U	U	U

Table 14.1. (Continued)

Corrosive	25°C	65°C	90°C
Perchloric acid	Q	U	U
Phenol	U	U	U
Phosphoric acid	S	S	S
Phthalic acid	S	S	S
Phthalic anhydride	S	Q	Q
Potassium hydroxide	S	S	S
Potassium permanganate	Q	Q	U
Potassium salts	S	S	S
Pyridine	U	U	U
Silver salts	S	S	S
Soap	S	S	S
Sodium salts	S	S	S
Sodium chlorate	Q	U	U
Sodium hydroxide	S	S	S
Sodium hypochlorite	Q	U	U
Sulfur dioxide	S	S	S
Sulfur trioxide	Q	Q	U
Sulfuric acid (100%)	U	U	U
Sulfuric acid (70%)	S	S	Q
Sulfuric acid (1–50%)	S	S	S
Tetralin	U	U	U
Tin salts	S	S	S
Toluene	U	U	U
Trichloroethylene	U	U	U
Turpentine	U	U	U
Urine	S	S	S
Vegetable oils	U	U	U
Wine	S	S	S
Xylene	U	U	U
Zinc salts	S	S	S

[a]Legend: S = satisfactory; Q = questionable; U = unsatisfactory.

[b]Useful information may be obtained for other corrosives by referring to data on related materials such as vinegar (dilute acetic acid), cleaning solvents (trichloroethylene), acetate solvents (amyl acetate), corn oil (vegetable oil), lubricating oil (mineral oil), aromatic hydrocarbons (benzene), etc.

Table 14.2. Listing of trade name or brand name, product characteristics, and manufacturer

Trade Name or Brand Name	Product	Manufacturer
Abafil	Reinforced ABS	Rexall Chemical Co.
Abcite	Plastic sheet	E. I. duPont de Nemours & Co., Inc.
ABS-213	ABS plastic	Dow Chemical Co.
Absafil G-1200/20	Glass-filled ABS plastic	Fiberfil/Dart
Abson	ABS polymers	B. F. Goodrich Chemical Co.
Abson 89161	ABS plastic	Mobay Chemical Co.
Aclar	Polyfluorocarbon film	Allied Chemical Corp.
Acralen	Styrene–butadiene latex	Farbenfabriken Bayer AG
Acrivue 360	Acrylic sheet	Swedlow, Inc.
Acronal	Polyalkyl vinyl ether	General Aniline & Film Corp.
Acrylacon	Fibrous-glass-reinforced polymers	Rexall Chemical Co.
Acrylafil	Reinforced polymers	Rexall Chemical Co.
Acrylafil G-40/20	Fibrous-glass-reinforced SAN plastic	Fiberfil/Dart
Acrilan	Polyacrylonitrile	Chemstrand Co.
Acrylite	Polymethyl methacrylate	American Cyanamide Co.
Acrylivin	Acrylic PVC sheet	General Tire & Rubber Co.
Acryloid	Resins solution	Rohm & Haas Co.
Actol	Polyethers	Allied Chemical Corp.
Acu-pak AP 8000	Adhesive (RT-cure)	Fenwal Inc.
Adiprene	Urethane elastomers and prepolymers	E. I. duPont de Nemours & Co., Inc.
Adiprene BL-16	Polyurethane elastomer liquid	E. I. duPont de Nemours & Co., Inc.
Aerodux	Resorcinol-formaldehyde resin	Ciba (A.R.L.) Ltd.
Alathon	Polyethylene	E. I. duPont de Nemours & Co., Inc.
Alathon 1540	Polyethylene plastic	E. I. duPont de Nemours & Co., Inc.
Albertols	Phenolic resins	
Aldocryl	Acetal resin	Shell Chemical Co.
Alfane	Epoxy resin cement	The Atlas Mineral Products Co.
Alkathene	Polyethylene resins	Imperial Chemical Industries, Ltd.
Alkathene 11	Polyethylene plastic	Imperial Chemical Industries, Ltd.
Alkon	Acetal copolymer	Imperial Chemical Industries, Ltd.
Alkor	Furan resin cement	Atlas Mineral Products Co.
Alloprene	Chlorinated natural rubber	Imperial Chemical Industries, Ltd.
Alsynite	Reinforced plastic panel	Reichhold Chemicals, Inc.
Amberlac	Modified alkyd resins	Rohm & Haas Co.
Amberol	Phenolic resins	Rohm & Haas Co.
Amerhoechst QCA-2460	Polyvinyl chloride film	American Hoechst
Ameripol	Polyethylene	Goodrich-Gulf Chemical Inc.
Ameripol CB 220	Elastomer-gum stk butadiene	B. F. Goodrich Chemical Co.
Ameripol 1001	Elastomer-gum stk styrene butadiene	B. F. Goodrich Chemical Co.
Amerith	Cellulose nitrate	Celanese Corp. of America
Amerlite	Ion-exchange resins	Rohm & Haas Co.
Amidel TN-1540	Nylon	Union Carbide
Amilan	Nylon	Tojo Rayon Co.
AMOCO Grade H5M	Polystyrene	Amoco Chemical Corp.
Ampcoflex	Rigid polyvinyl chloride	Atlas Mineral Products Co.
Amsco Melt 107	Ethylene vinyl-acetate adhesive	Union Oil of California
Aquaflex OR-30	Polyvinyl chloride	Pantasote
Araidite AT1	Two-part (RT-cure) epoxy	Ciba Geigy

Table 14.2. (Continued)

Trade Name or Brand Name	Product	Manufacturer
Araidite AV121B/HY930	Epoxy adhesive	Ciba Geigy
Araldite	Epoxy resins	CIBA (A.R.L.) Ltd.
Arcon ER-535	Polyurethane (TS) adhesive	Allied Resin Corp.
Armco 73201 SMC	Polyester sheet	Armco
Armite	Vulcanized fibre	Spaulding Fibre Co.
Armstrong A-12T	Epoxy	Armstrong
Arnel	Cellulose triacetate	Celanese Corp. of America
Arnite	Polyethylene terephthalate	Algemene Kunstzijde Uni N.V.
Arnite A04 900	Glass-fiber-filled polyethylene terephthalate	Akzo Plastics
Arnox 1000	Mineral-filled epoxy	General Electric Co.
Arofene	Phenolic resins	Ashland Chemical Co.
Aropol	Polyester resins	Ashland Chemical Co.
Aropol 7320	Glass-filled polyester	Ashland Chemical Co.
Aroset 1044-Z-40	Acrylic adhesive	Ashland Chemical Co.
Arothane	Polyester resins	Ashland Chemical Co.
ASRC 1013	Styrene–butadiene elastomer-gum	American Synthetic
Astrel Co.	Polyarylsulfone	3M Co.
Atlac 580E 05A	Glass-laminate vinyl esters	ICI Americas Inc.
Atlac 711-05A	Glass-laminate polyester	ICI Americas Inc.
Atlas E-21	Epoxy two-part adhesive	Oak Materials Grp.
Avron	Rayon fiber	American Viscose Corp.
Azdel	Fibrous-glass-reinforced ABS copolymer sheet	Generic Name
Aztran	Poromeric sheet	B. F. Goodrich Chemical Co.
Bakelite	Phenol-formaldehyde	Union Carbide Chemical Co.
Bakelite CO-merDPD 6169	Ethylene-ethylacrylate	Union Carbide Chemical Co.
Bakelite DFDA0140 Nat	Polyethylene plastic	Union Carbide Chemical Co.
Bakelite DHDA6444 Bik	Ethylene-vinylacetate	Union Carbide Chemical Co.
Bakelite DPDA2304 Nat	Ethylene-ethylacrylate	Union Carbide Chemical Co.
Bakelite ERL ERR	Cycloaliphatic epoxide	Union Carbide Chemical Co.
Bakelite H34FR	Glass fabric epoxy laminate	Bakelite Xylonite, Ltd.
Barex	Barrier resin	Vistron Corp.
Bayblend MC-2500	Polycarbonate plastic	Mobay Chemical Co.
Baygal	Polyester for casting resins	Farbenfabriken Bayer AG
Baylon V 10 H 460	Polyethylene plastic	Bayer AG
Baypren	Polychloroprene	Farbenfabriken Bayer AG
Beckacite	Modified phenolic resin	Reichhold Chemicals, Inc., and Beck, Koller & Co., Ltd.
Beckamine	Urea-formaldehyde resin	Reichhold Chemicals, Inc., and Beck, Koller & Co., Ltd.
Beckosol	Alkyd resins	Reichhold Chemicals, Inc., and Beck, Koller & Co., Ltd.
Beetle	Urea-formaldehyde resins	American Cyanamid Co.
Benvic	Polyvinyl chloride	Solvay & Cie, S.A.
Bexphane	Polypropylene	Bakelite Xylonite, Ltd.
BFG A-1232-B	Epoxy	B. F. Goodrich Chemical Co.
BFG A-203-B	Chloroprene adhesive	B. F. Goodrich Chemical Co.
Bicor EU	Polypropylene film	Mobil Chemical Co.
Blendex	ABS resin	Borg-Warner Corp.
Bolta Flex	Vinyl sheeting and film	General Tire & Rubber Co.
Boltaron	Plastic sheets	General Tire & Rubber Co.

Table 14.2. (Continued)

Trade Name or Brand Name	Product	Manufacturer
Boltaron 1013-9019 PC	Polyvinyl chloride	General Tire & Rubber Co.
Boltaron 5000	Polyethylene sheet	General Tire & Rubber Co.
Boltaron 6150	ABS sheet	General Tire & Rubber Co.
Bondmaster 82-0796/97	Epoxy adhesive	National Starch
Bondstrand	Filament-wound fiber-glass reinforced plastics	Ameron Corrosion Control Div.
Borden 3045	Polyvinyl chloride plastic	Borden Inc.
Boronol	Polyolefins with boron	Allied Resinous Products, Inc.
Bostik M890/A	Acrylic adhesive	Bostik
Bostik 1036	Chloroprene adhesive	Bostik
Bostik 7087	Epoxy adhesive	Bostik
Bostik 7696	Polyester adhesive	Bostik
Budene 1207	Butadiene elastomer-gum	Goodyear Chemicals
Butacite	Polyvinyl acetal resins	E. I. duPont de Nemours & Co., Inc.
Butakon	Butadiene copolymers	Imperial Chemical Industries, Ltd.
Butaprene	Styrene–butadiene elastomers	Firestone Tire & Rubber Co.
Butarez CTL	Telechelic butadiene polymer	Phillips Petroleum Co.
Buton	Butadiene–styrene resin	Enjay Chemical Co.
Bu-Tuf	Polybutene	Petrotex Chemical Corp.
Butvar	Polyvinyl butyral resin	Shawinigan Resins Corp.
Cadco Delrin AF	Acetal sheet	Cadillac Plastics
Cadco FEP film	Fluorocarbon film	Cadillac Plastics
Cadco GPO-1	Polyester FRP	Cadillac Plastics
Cadco Nylon-6/6	Nylon-6/6 sheet	Cadillac Plastics
Capran	Nylon-6	Allied Chemical Corp.
Capran Deepform	Nylon-6 film	Allied Chemical Corp.
Caprez DPP	Polypropylene plastic	MPD Technology
Carbomastic	Epoxy coal-tar coating	Carboline Co.
Carboset	Acrylic resins	B. F. Goodrich Chemical Co.
Cariflex I	Cis-1,4-polyisoprene	Shell Chemical Co., Ltd.
Carina	Polyvinyl chloride	Shell Chemical Co., Ltd.
Carinex	Polystyrene	Shell Chemical Co., Ltd.
Castethane	Castable polyurethanes	Upjohn Co.
Castethane CPR2118-83A	Polyurethane polyester	Upjohn Co.
Catalac	Phenol-formaldehyde resin	Catalin Ltd.
Ceilcote P-370	Polyester primer	Ceilcote Co.
Ceilcote 680 Primer/Sealer	Epoxy	Ceilcote Co.
Celanar	Polyester film and sheeting	Celanese Plastics Co.
Celanese LA 201M	Acrylic pressure-sensitive adhesive	Celanese Polymers
Celanese 1003	Nylon-6/6	Celanese Plastics Co.
Celanex	Thermoplastic polyester	Celanese Plastics Co.
Celanex 2000	Polyester (PBT) plastics	Celanese Plastics Co.
Celcon	Acetal copolymers	Celanese Plastics Co.
Celcon M140	Acetal copolymer	Celanese Plastics Co.
Celgard	Microporous polypropylene	Celanese Plastics Co.
Celgard 2402	Polypropylene film	Celanese Plastics Co.
Cellofoam	Polystyrene-foam board	United States Mineral Products Co.
Celluloid	Plasticized cellulose nitrate	Celanese Plastics Co.
Cerex	Styrene copolymer	Monsanto Chemical Co.
Chemigum	Urethane elastomer	Goodyear Tire & Rubber Co.
Chemigum N328B	Acrylonitrile–butadiene elastomer-gum stk	Goodyear Chemicals

Table 14.2. (Continued)

Trade Name or Brand Name	Product	Manufacturer
Chem-O-Sol	PVC plastisol	Chemical Products Co.
Chemplex 1005	Polyethylene plastics	Chemplex
Cibanite	Aniline-formaldehyde resin	Ciba Products Co.
Cis-4	Cis-1,4-polybutadiene	Phillips Petroleum Co.
Clearseal K2	Polyvinyl chloride film	Columbus Coated
Cleartuf 1002	Polyethylene terephthalate	Goodyear Chemicals
Clysar ECL	Polyethylene film	E. I. duPont de Nemours & Co., Inc.
Co-Cel	Nitrocellulose film	St. Regis Paper Co.
Cold Microblack 1605	Styrene–butadiene elastomer-gum stk	B. F. Goodrich Chemical Co.
Columbus K6Lt	Polyvinyl chloride	Columbus Coated
Conoco 14331	Polyvinyl chloride	Conoco Chemicals
Coral Rubber	Cis-polyisoprene	Firestone Tire & Rubber Co.
Cordo	PVC foam and films	Ferro Corp.
Corfam	Poromeric film	E. I. duPont de Nemours & Co., Inc.
Coriar 5150	Glass-fabric epoxy laminate	E. I. duPont de Nemours & Co., Inc.
Corogard 9	Unspecified polymer	3M Co.
Corvel	Plastic coating powders	The Polymer Corp.
Corvic	Vinyl polymers	Imperial Chemical Industries, Ltd.
Cosden 500	Polystyrene plastic	Cosden Oil & Chemicals
CPR-9006-4	Polyurethane foam-semifabric	Upjohn Co.
Creslan	Acrylonitrile-acrylic ester copolymers	American Cyanamid Co.
Cronar	Polyethylene	E. I. duPont de Nemours & Co., Inc.
Cryorap	Thermoplastic sheets	W. R. Grace & Co.
Cryovac H.S. 660	Polyolefin film	Cryovac Div.
Crystalex	Acrylic resin	Rohm & Haas Co.
Crystic	Polyester resins	Scott Bader Co.
CS 725	Polyethylene copolymer	Cities Service Co.
Cumar	Coumarone indene resins	Allied Chemical Corp.
Cyanacryl C	Acrylic elastomer	American Cyanamid Co.
Cyanaprene	Polyurethane	American Cyanamid Co.
Cyanaprene A-7	Polyurethane	American Cyanamid Co.
Cyanobone CB-P01	Cyanoacrylate	Sumitomo Chem. Co.
Cybond 4533 RX	One-part epoxy adhesive	American Cyanamid Co.
Cyclass 2693	Glass-filled polyester	American Cyanamid Co.
Cycolac	Acrylonitrile–butadiene–styrene copolymer	Borg-Warner Corp.
Cycolac CG	ABS plastic	Borg-Warner Corp.
Cymac	Thermoplastic molding materials	American Cyanamid Co.
Cymel	Melamine molding compound	American Cyanamid Co.
C15-024	Unfilled epoxy	Hysol/Dexter
Dacovin	Rigid polyvinyl chloride	Diamond Alkali Co.
Dacovin CRS-704	Polyvinyl chloride	Diamond Shamrock
DAP One-Part Flexiseal	Polysulfide sealant	DAP, Inc.
DAPCO Rollstock	Polyethylene film	DAPCO
Dapon	Diallyl-phthalate prepolymer	FMC Corp.
Daponite	Dapon-fabric laminates	FMC Corp.
Daran	Polyvinylidene chloride	W. R. Grace & Co.
Daratak A	Polyvinyl acetate emulsion adhesive	W. R. Grace & Co.
Darex	Styrene copolymer resin	W. R. Grace & Co.

Table 14.2. (Continued)

Trade Name or Brand Name	Product	Manufacturer
Darvan	Polyvinylidene cyanide	Celanese Corp. of America
Darvic	Polyvinyl chloride	Imperial Chemical Industries, Ltd.
Daubond DC 8608	Epoxy adhesive	Daubert Chemical
Delrin	Acetal polymer	E. I. duPont de Nemours & Co., Inc.
Delrin AF113	Acetal polymer 21% TFE	E. I. duPont de Nemours & Co., Inc.
DEN 331	Epoxy plastic	Dow Chemical Co.
DER 330	Epoxy plastic	Dow Chemical Co.
Derakane	Polyester resin (vinyl ester)	Dow Chemical Co.
Derakane 411	Vinyl ester, monomeric styrene (45%)	Dow Chemical Co.
Desmocoll E-361	Polyurethane (TS)	Mobay Chemical Co.
Desmopan	Polyurethanes	Farbenfabriken Bayer AG
Devcon Clear Epoxy	Epoxy two-part (RT-cure)	Devcon Corp.
Devran	Epoxy resins	Devoe & Reynolds Co.
Dexel	Cellulose acetate	British Celanese Ltd.
Dexsil	Polycarboranesiloxane	Olin Mathieson Corp.
Diakin	Polymethyl methacrylate	Imperial Chemical Industries, Ltd.
Diakon	Acrylic	Imperial Chemical Industries, Ltd.
Diall	Diallyl phthalate	Allied Chemical Corp.
Diene	Polybutadiene	Firestone Tire & Rubber Co.
Dimetcote	Protective coating	Americoat Corp.
Dion	Polyester resin	Diamond Alkali Co.
Dow Corning Bathtub Caulk	Silicone sealant	Dow Corning Corp.
Dow CPE-CM 0136	Chlorinated polyethylene	Dow Chemical Co.
Dow HB 12365	Polyethylene	Dow Chemical Co.
Dow Silicone Rubber Seal	Silicone sealant	Dow Corning Corp.
DPR 242-H Part A	Isoprene elastomer liquid	Hardman, Inc.
Dualc	Lacquers	Sun Chemical Corp.
Duco	Cellulose nitrate lacquers	E. I. duPont de Nemours & Co., Inc.
Dulux	Polymeric enamels	E. I. duPont de Nemours & Co., Inc.
DuPont NR-056X	Polyimide adhesive	E. I. duPont de Nemours & Co., Inc.
DuPont 46944	Polyester adhesive	E. I. duPont de Nemours & Co., Inc.
DuPont 68070	Acrylic adhesive	E. I. duPont de Nemours & Co., Inc.
Durabond Acrylic Caulk	Acrylic sealant	U.S. Gypsum Co.
Duragen 1203	Butadiene elastomer	General Tire & Rubber Co.
Duralon	Furan molding resins	U.S. Stoneware Co.
Duramac	Alkyd resins	Commercial Solvents Corp.
Duraplex	Alkyd resins	Rohm & Haas Co.
Durathon	Polybutylene resins	Witco Chemical Corp.
Durethane	Nylon-6	Farbenfabriken Bayer AG
Durethan U	Polyurethanes	Farbenfabriken Bayer AG
Durethene	Polyethylene film	Sinclair-Koppers Co., Inc.
Durez 118 black	Filled phenolic plastic	Hooker Chemicals & Plastics
Durez 22761 green	Glass-fiber diallyl phthalate plastic	Hooker Chemicals & Plastics
Durez 24060 black	Glass-alkyd plastic	Hooker Chemicals & Plastics
Durez 24776 green	Glass-fiber diallyl phthalate plastic	Hooker Chemicals & Plastics
Durite	Phenolic resins	The Borden Co.
Duro-Flex 25 30-9125	Polyurethane (TS) two-part adhesive	National Starch
Dyal	Alkyd resins	Sherwin-Williams Co.
Dylan	Polyethylene resins	Sinclair-Koppers Co., Inc.

Table 14.2. (Continued)

Trade Name or Brand Name	Product	Manufacturer
Dylan B-375	Polyethylene	ARCO/Polymer, Inc.
Dylark 232	Polystyrene	ARCO/Polymer, Inc.
Dylel	ABS copolymer	Sinclair-Koppers Co., Inc.
Dylene	Polystyrene resins	ARCO/Polymer, Inc.
Dylite	Expandable polystyrene	Sinclair-Koppers Co., Inc.
Dynafilm	Polypropylene film	U.S. Industrial Chemicals Co., Div. National Distillers & Chemical Corp.
Dynel	Modacrylic fiber	Union Carbide Corp.
Dyphene	Phenol-formaldehyde resins	Sherwin-Williams Co.
Dypro K-122	Ethylene–propylene copolymer	ARCO/Polymer, Inc.
E-21718	Carbon-graphite epoxy	Fiberite Corp.
Eastman CHDM	Polyol and isocyanate glycol	Eastman Chemical Products, Inc.
Eccobond Paste E	Epoxy adhesive	Emerson & Cuming
Eccobond Solder 59C	Silicone adhesive	Emerson & Cuming
Eccobond 300	Acrylic adhesive	Emerson & Cuming
Eccofoam FP	Polyurethane cast/encap/potting	Emerson & Cuming
Eccosil DC 23 RTV	Silicone elastomer/liquid	Emerson & Cuming
Edistir AR	Polystyrene plastic	Montedison
EF 406	Polyethylene	Cities Service Co.
Elastothane 625M	Polyurethane elastomer	Thiokol Chemical Co.
Elexar 8421Z	Styrene polyolefin	Shell Chemical Co.
El Rexene	Polyolefin resins	Rexall Chemical Co.
El Rey	Low-density polyethylene	Rexall Chemical Co.
Elvacet	Polyvinyl acetate emulsion	E. I. duPont de Nemours & Co., Inc.
Elvacite	Acrylic resins	E. I. duPont de Nemours & Co., Inc.
Elvamide 8023	Nylon adhesive	E. I. duPont de Nemours & Co., Inc.
Elvanol	Polyvinyl alcohol resins	E. I. duPont de Nemours & Co., Inc.
Elvax	Polyethylene-co-vinyl acetate	E. I. duPont de Nemours & Co., Inc.
Elvax 150	Ethylene vinylacetate	E. I. duPont de Nemours & Co., Inc.
EM 447	Polyethylene	Cities Service Co.
Enrad	Preirradiated polyethylene	Enflo Corp.
Enrup	Thermosetting resin	United States Rubber Co.
Ensolite AA	PVC/nitrile foam-semifabric	Uniroyal, Inc.
Epcar 346	EPDM elastomer	B. F. Goodrich Chemical Co.
Epibond	Epoxy adhesive resin	Furane Plastics, Inc.
Epikote	Epoxy resins	Shell Chemical Co.
Epi-Rez	Epoxy cast resin	Celanese Corp.
EPI-WELD Type II	Epoxy adhesive	Allied Composition
Epibond 104 A/B	Unspecified polymer	M & T Chemical
Epocast	Epoxy resins	Furane Plastics, Inc.
Epocryl	Epoxy acrylate resin	Shell Chemical Co.
Epodite	Epoxy resins	Showa Highpolymer Co.
Epolene	Low-melt polyethylene	Eastman Chemical Products, Inc.
Epon	Epoxy resins	Shell Chemical Co.
Epotuf	Epoxy resins	Reichhold Chemical Co.
Epoweld 3243	Epoxy adhesive	Hardman, Inc.
Epoxylite	Epoxy resins	Epoxylite Corp.
Epoxylite R-69/CatN	Epoxy plastic	Epoxylite Corp.
Epsyn 3506	EPDM elastomer	Copolymer Rubber
Escurene	Polyethylene	Exxon Chemical Co., U.S.A.
Estane	Polyurethane resins	B. F. Goodrich Chemical Co.
Estane 58013	Polyurethane elastomer TPR	B. F. Goodrich Chemical Co.

Table 14.2. (Continued)

Trade Name or Brand Name	Product	Manufacturer
Ethafoam-SB	Polyethylene LDPE	Foamade Ind.
Ethocel	Ethylcellulose	Dow Chemical Co.
Ethyl 5009	Polyvinyl chloride	Ethyl Corp.
EVA 3120	Ethylene vinylacetate	E. I. duPont de Nemours & Co., Inc.
Evenglo	Polystyrene	Sinclair-Koppers Co., Inc.
Exon	Polyvinyl chloride	Firestone Plastics
Extrel	Plastic films	Exxon Chemical Co., U.S.A.
Extrel EBW-2	Polyethylene film	Exxon Chemical Co., U.S.A.
Exxon CD-119D	Polypropylene plastic	Exxon Chemical Co., U.S.A.
FA High Viscous Polysulfide	Polysulfide elastomer	Thiokol Chemical Co.
Fabrikoid	Pyroxylin-coated fabrics	E. I. duPont de Nemours & Co., Inc.
Fastbond 10	Chloroprene contact	3M Co.
Fibercast	Reinforced plastic pipe	Fibercast Co.
Fiberite	Phenolic molding compounds	Fiberite Corp.
Firestone Foam F-1833	Polyurethane foam-semifabric	Firestone Tire & Rubber Co.
Flamolin XL 7400	Flame retarded polyolefin	U.S. Industrial Chemical
Flexiseal Clear SW-100	Synthetic rubber sealant	DAP, Inc.
Fluon	Polytetrafluoroethylene	Imperial Chemical Industries, Ltd.
Fluon G163	PTFE fluorocarbon	Imperial Chemical Industries, Ltd.
Fluorel	Polyvinylidene fluoride	3M Co.
Fluorobestos	Asbestos-teflon composite	Raybestos Manhatten, Inc.
Fluoron	Polychlorotrifluoroethylene	Stokes Molded Products
Fluoroplast	Polytetrafluoroethylene	U.S. Gasket Co.
FM 10365	Cellulose-filled phenolic plastic	Fiberite Corp.
FM 3000	Cellulose-filled phenolic plastic	Fiberite Corp.
FM—53U	Unsupported epoxy film	American Cyanamid Co.
Foamex	Polyvinyl formaldehyde	General Electric Co.
Formaldafil AC-80/TF/SS	PTFE acetal plastic	Fiberfil/Dart
Formex	Polyvinyl acetal	General Electric Co.
Formica	Thermosetting laminates	Formica Corp.
Formvar	Polyvinyl formal	Shawinigan Resins Corp.
Forticel	Cellulose propionate	Celanese Corp. of America
Fortiflex	Polyethylene	Celanese Plastics Co.
Fortiflex B45-30H-02	Carbon black-filled polyethylene	Soltex Polymer Co.
Fortrel	Polyester fiber	Fiber Industries, Inc.
Fostacryl	Polystyrene-co-acrylonitrile	Foster Grant Co.
Fostalite 879	Polystyrene	American Hoescht
Fostanylon 1047	Nylon-6	American Hoescht
Fostarene	Polystyrene	Foster Grant Co.
FPC	PVC resins compound	Firestone Tire & Rubber Co.
FPC 1238	Polyvinyl chloride	Firestone Tire & Rubber Co.
FR-S 146	Styrene–butadiene elastomer	Firestone Tire & Rubber Co.
FRS 1502	Styrene–butadiene elastomer	Firestone Tire & Rubber Co.
Fulton 404	PTFE acetal plastic	LNP Corp.
Furnane	Epoxy and furan resins	Atlas Mineral Products Co.
Galaith	Casein plastics	
Gantrez	Polyvinyl ether-co-maleic anhydride	Dyestuff & Chemical Div., General Aniline & Film Corp.
GE RTV 102	Silicone sealant	General Electric Corp.
Gelva	Polyvinyl acetate	Shawinigan Resins Corp.
Gelva GMS-263	Pressure-sensitive acrylic adhesive	Monsanto Co.
Gelvatex	Polyvinyl acetate emulsions	Shawinigan Resins Corp.

Table 14.2. (Continued)

Trade Name or Brand Name	Product	Manufacturer
Gelvatol	Polyvinyl alcohol	Shawinigan Resins Corp.
Genaire	Poromeric film	General Tire & Rubber Co.
Genal	Thermosets	General Electric Co.
Genal GF 7011	Glass-fiber phenolic plastic	General Electric Co.
General Tire BF5-3896	Polyvinyl chloride film	General Tire & Rubber Co.
Genetron	Fluorinated hydrocarbon monomers and polymers	Allied Chemical Co.
Genthane	Polyurethane elastomer	General Tire & Rubber Co.
Gentro	Butadiene copolymer	General Tire & Rubber Co.
Gentro-Jet 1601 Mstrbtch	Polyvinyl chloride	American Hoescht
Gentro-Jet 9106	Styrene–butadiene	General Tire & Rubber Co.
Geon	Polyvinyl chloride	B. F. Goodrich Chemical Co.
Geon 2042	Styrene–butadiene elastomer	B. F. Goodrich Chemical Co.
GL-200	Acrylic copolymer	Richardson Co.
Glaskyd	Glass-reinforced alkyd resin	American Cyanamid Co.
Glaskyd 1901	Glass-reinforced alkyd	American Cyanamid Co.
Glyptal	Alkyd coating	General Electric Co.
Goodyear B17	Polyvinyl chloride film	Goodyear Tire Co.
Grex	Polyethylene	W. R. Grace & Co.
Gulf 6210MJ	Polypropylene	Gulf Oil Chemical
H.P.P. 80	Polypropylene	Cole Plastics Ltd.
Halar	Polyfluorocarbons	Allied Chemical Corp.
Halar ECTFE	Fluoropolymer	Allied Chemical Corp.
Halex	Polyfluorocarbon	Allied Chemical Corp.
Halon	Fluorochlorocarbon	Allied Chemical Corp.
Halowax	Chlorinated naphthalene	Union Carbide Corp.
Hardman DC-2900-HW	Epoxy cast/encap. potting	Hardman Inc.
Hathane 1640C-12 Pt.A	Polyurethane foam-liquid	Hastings
Havamold HM 9100 SMC	Glass-fiber-filled polyester sheet	Haveg Industries, Inc.
HB Fuller SC-1189	Chloroprene sealant	H. B. Fuller Co.
Herclor C	Epi-cl-hydrin copolymer	Hercules Powder Co.
Hercose	Cellulose acetate-propionate	Hercules Powder Co.
Hercules N-type	Ethyl cellulose	Hercules Powder Co.
Herculoid	Cellulose nitrate	Hercules Powder Co.
Herculon	Polypropylene	Hercules Powder Co.
Herox	Nylon	E. I. duPont de Nemours & Co., Inc.
Hetron	Fire-retardant polyester resins	Hooker Chemical Corp.
Heveaplus	Copolymer of methyl-methacrylate and rubber	
H-film	Polyimide film	E. I. duPont de Nemours & Co., Inc.
Hi-Blen	ABS polymers	Japanese Geon Co.
Hi-D 6420	Polyethylene	Gulf Oil Chemicals
Hi-fax	High-density polyethylene	FMC Corp. and Hercules Powder Co.
Hipack	Polyethylene	Showa Highpolymer Co.
Hitalex	Polyethylene	Hitachi Chemical Co.
Hitanol	Phenol-formaldehyde resins	Hitachi Chemical Co.
Hostaflon C2	Polychlorotrifluoroethylene	Farbwerke Hoechst AG
Hostaflon TF	Polytetrafluoroethylene	Farbwerke Hoechst AG
Hostalen	Polyethylene	Farbwerke Hoechst AG
Hostalen GF 4750	Polyethylene	American Hoechst
Hybond C-9175A	Chloroprene contact	Pierce & Stevens

Table 14.2. (Continued)

Trade Name or Brand Name	Product	Manufacturer
Hycar	Butadiene–acrylonitrile	B. F. Goodrich Chemical Co.
Hycar Nitrile 1031	Acrylonitrile–butadiene elastomer	B. F. Goodrich Chemical Co.
Hycar 4021(PA-21)	Acrylic elastomer	B. F. Goodrich Chemical Co.
Hydrophylic-1 hydrophylic	Polyurethane foam-semifabric	Tenneco Chemical
Hydropol	Hydrogenated polybutadiene	
Hypalon	Chlorosulfonated-polyethylene	E. I. duPont de Nemours & Co., Inc.
Implex	Acrylic resins	Rohm & Haas Co.
Implex A	Acrylic	Rohm & Haas Co.
Insurok	Phenol-formaldehyde molding compounds	The Richardson Co.
Insurok T-160	Phenolic-paper-laminate	Spaulding Fibre
Intamix	Rigid PVC	Diamond Shamrock Corp.
Irrathene	Irradiated polyethylene	General Electric Co.
Isofoam FF-2	Polyurethane foam-liquid	Witco Chemical Corp.
Isomid	Polyester-polyimide film magnetic wire	Schenectady Chemicals, Inc.
Isonate CPR 402	Polyurethane foam-liquid	Upjohn Co.
Iupilon	Polycarbonate	Mitsubishi Edogawa Chemical Co.
J-Thane 63B	Polyurethane (TS)	Jefferson Chemical Co.
Jet-Kote	Furane resin coatings	Furane Plastics Inc.
Kapton	Polyimide	E. I. duPont de Nemours & Co., Inc.
Kardel	Polystyrene film	Union Carbide Corp.
Kaurit	Urea-formaldehyde resins	Badische Anilin & Soda-Fabrik AG
Kel-F	Trifluorochloroethylene resins	3M Co.
Keltrol	Copolymers	Textron, Inc.
Kematal	Acetal copolymers	Imperial Chemical Industries, Ltd.
Kenflex	Hydrocarbon resins	Kenrich Petrochemicals, Inc.
Ketac	Ketone-aldehyde resins	American Cyanamid Co.
Kodacel	Cellulose acetate film	Eastman Chemical Products, Inc.
Kodar A150	Polyester (copolyester)	Eastman Chemical Products, Inc.
Kollidon	Polyvinyl pyrrolidone	General Aniline & Film Corp.
Kopox	Epoxy resin	Koppers Co.
Koppers S3801-25	Polyester	Koppers Co.
Korad	Acrylic film	Rohm & Haas Co.
Korex	Phenolic resin cement	Atlas Mineral Products Company
Koroseal	Polyvinyl chloride	B. F. Goodrich Chemical Co.
Koylon-Latex Foam-1	Chloroprene foam-semifabric	Uniroyal, Inc.
Kralac	ABS resins	Uniroyal, Inc.
Kralastic	ABS	Uniroyal, Inc.
Kralastic FVM	ABS/PVC	USS Chemicals
Kralon	High-impact styrene and ABS resins	Uniroyal, Inc.
Kraton	Butadiene copolymers	Shell Chemical Co.
Kraton G 1650	Styrene-ethylene elastomer	Shell Chemical Co.
Krene	Plasticized vinyl film	Union Carbide Corp.
K-Resin	Butadiene–styrene copolymer	Phillips Petroleum Co.
K-Resin KR01	Styrene–butadiene copolymer	Phillips Petroleum Co.
Kriston	Allyl ester casting resins	B. F. Goodrich Chemical Co.
Kydex	Acrylic-polyvinylchloride sheet	Rohm & Haas Co.
Kydex 100	Acrylic/PVC	Rohm & Haas Co.
Kynar	Polyvinylidene fluoride	Pennwalt Chemicals Corp.
Kynar 316	Polyvinylidene fluoride	Pennwalt Chemicals Corp.

Table 14.2. (Continued)

Trade Name or Brand Name	Product	Manufacturer
Laminac	Polyester resins	American Cyanamid Co.
Leguval	Polyester resins	Farbenfabriken Bayer AG
Lekutherm	Epoxy resins	Farbenfabriken Bayer AG
Lemac	Polyvinyl acetate	Borden Chemical Co.
Lemol	Polyvinyl alcohol	Borden Chemical Co.
Levapren	Ethylene–vinylacetate copolymers	Farbenfabriken Bayer AG
Lexan	Polycarbonate resins	General Electric Co.
Lexan CTG	Polycarbonate	General Electric Co.
Lock Foam	Polyurethane foam	Nopco Chemical Co.
Loctite Adh/Sealant 277	Methacrylate anaerobic sealant	Loctite Corp. USA
LR601A Compound	Polyurethane elastomer	Phillips Chemical Co.
Lucite	Polymethyl methacrylate	E. I. duPont de Nemours & Co., Inc.
Lucite L	Acrylic sheet	E. I. duPont de Nemours & Co., Inc.
Lumarith	Cellulose acetate	Celanese Corp. of America
Lustran	Molding and extrusion resins	Monsanto Chemical Co.
Lustran ABS HR-850	ABS	Monsanto Chemical Co.
Lustran SAN 31	SAN copolymer	Monsanto Chemical Co.
Lustrex	Polystyrene	Monsanto Chemical Co.
Lustrex HF 555	Polystyrene	Monsanto Chemical Co.
Lutonal	Polyvinyl ethers	Badische Anilin & Soda-Fabrik AG
Lutrex	Polyvinyl acetate	Foster Grant Co.
Luvican	Polyvinyl carbazole	Badische Anilin & Soda-Fabrik AG
MAAS	Flame-retarded polystyrene	Cole Plastics Ltd.
Magnamite 1908/AS4-108	Graphite epoxy	Hercules Powder Co.
Makrolon	Polycarbonate	Farbenfabriken Bayer AG
Makrotol	Polycarbonate film	Naftone, Inc.
Maranyl	Nylons	Imperial Chemical Industries, Ltd.
Maraset	Epoxy resins	The Marblette Corp.
Marbon	Polystyrene and copolymers	Borg-Warner Corp.
Markel Econosculp	Polyurethane foam-semifabrics	Market Rubber Products
Marlex	Polyolefin resins	Phillips Chemical Co.
Marvibond	Metal-plastics laminates	Uniroyal, Inc.
Marvinol	Polyvinyl chloride	Uniroyal, Inc.
Maxhesive 25	Polyester adhesive	Ceilcote Co.
Melan	Melamine resins	Hitachi Chemical Co., Ltd.
Melinex	Polyethylene terephthalate	Imperial Chemical Industries, Ltd.
Melinex AS	Polyethylene terephthalate copolymers	Imperial Chemical Industries, Ltd.
Melit	Melamine–formaldehyde resins	Societa Italiana Resine
Melmac	Melamine molding materials	American Cyanamid Co.
Melurac	Melamine-urea resins	American Cyanamid Co.
Merlon AL-400	Polycarbonate	Mobay Chemical Co.
Merlon	Polycarbonate	Mobay Chemical Co.
Methocel	Methylcellulose	Dow Chemical Co.
Metibond	Supported epoxy film	Narmco Materials
Micarta	Thermosetting laminates	Westinghouse Electric Corp.
Micarta H-12426	Glass-fabric-reinforced melamine formaldehyde	Westinghouse Electric Corp.
Microthene	Powdered polyethylene	U.S. Industrial Chemicals Co.
Microthene MA 778	Polyethylene	U.S. Industrial Chemical Co.
Minicel E-300	Polyethylene LDPE foam-semifabric	Voltek

Table 14.2. (Continued)

Trade Name or Brand Name	Product	Manufacturer
Minlon	Reinforced nylon	E. I. duPont de Nemours & Co., Inc.
Mirrex MCGP-1047	Chrome polyvinyl chloride	Tenneco Chemical
Mobay E-313	Isocyanate polyol and isocyanate	Mobay Chemical Co.
Mobile LBA-133	Polyethylene	Mobil Chemicals
Mobile 124 ESG	ABS	Mobil Chemicals
Modulene	Polyethylene resin	E. I. duPont de Nemours & Co., Inc.
Molpen	Polypropylene	Novamont Corp.
Moltopren	Polyurethane foam	Farbenfabriken Bayer AG
Mondur M-294	Isocyanate	Mobay Chemical Co.
Montrek	Polyethylene imine	Dow Chemical Co.
Moplen	Polypropylene	Montecatini Edison S.P.A.
Mowilith	Polyvinyl acetate	Farbwerke Hoechst AG
Mowiol	Polyvinyl alcohol	Farbwerke Hoechst AG
Mowital	Polyvinyl butyral	Farbwerke Hoechst AG
Multranol E-4011	Polyether polyol	Mobay Chemical Co.
Multron	Polyesters	Mobay Chemical Co.
MXB-5004 181-150s	Glass-fabric-reinforced polyimide	Fiberite Corp.
MXB-6001	Glass-fabric-reinforced phenolic laminated	Fiberite Corp.
MXB-7016	Glass-fabric-reinforced epoxy laminate	Fiberite Corp.
MXB-9002	Glass-fabric-reinforced polyester laminate	Fiberite Corp.
Mycalex	Inorganic molded plastics	Mycalex Corp of America
Mylar	Polyester film	E. I. duPont de Nemours & Co., Inc.
Mylar A	Polyester film	E. I. duPont de Nemours & Co., Inc.
Nalgon	Plasticized polyvinyl chloride	Nalge Co.
Napryl	Polypropylene	Pechiney-Saint-Gobain
Narmco 4065	Silica-reinforced phenolic laminate	Narmco Materials
Natene	Polyethylene	Pechiney-Saint-Gobain
Natsyn	Cis-(1,4 polyisoprene)	Goodyear Tire & Rubber Co.
Nebony	Petroleum hydrocarbon resin	Neville Chemical Co.
Neoprene	Polychloroprene	E. I. duPont de Nemours & Co., Inc.
Neoprene E-1	Chloroprene elastomer	Petro-Tex Chemicals
Neoprene TW	Chloroprene elastomer	E. I. duPont de Nemours & Co., Inc.
Nepoxide	Epoxy resin coating	Atlas Minerals & Chemicals Div., ESB, Inc.
Nevidene	Coumarone-indene resin	Neville Chemical Co.
Nevillac	Modified coumaron-indene resin	Neville Chemical Co.
Niax	Polyol polyesters	Union Carbide Corp.
Niax BDE-400	Polyether polyol	Union Carbide Corp.
Nimbus	Polyurethane foam	General Tire & Rubber Co.
Nipeon	Polyvinyl chloride	Japanese Geon Co.
Nipoflex	Ethylene–vinyl acetate copolymer	Toyo Soda Mfg. Co.
Nipolon	Polyethylene	Toyo Soda Mfg. Co.
Noan	Styrene–methyl methacrylate copolymer	Richardson Corp.
Nopcofoam	Polyurethane foams	Nopco Chemical Co.

Table 14.2. (Continued)

Trade Name or Brand Name	Product	Manufacturer
Nordel	Ethylene–propylene	E. I. duPont de Nemours & Co., Inc.
Nordel 1040	EPDM elastomer	E. I. duPont de Nemours & Co., Inc.
Noryl	Polyphenylene oxide	General Electric Co.
Noryl EN-185	Phenylene oxide	General Electric Co.
Novodur	ABS polymers	Farbenfabriken Bayer AG
Nuba	Modified coumarone	Neville Chemical Co.
Nuclon	Polycarbonate	Pittsburg Plate Glass Co.
Nukem	Acid-resistant resin cements	Amercoat Corp.
Ny-Kon R	Molysulfide nylon-6/6	LNP Corp.
Nylafil	Reinforced nylon	Rexall Chemical Co.
Nylafil G-1/30	Long-glass-reinforced nylon-6/6	Fiberfill/Dart
Nylasint	Sintered nylon parts	Polymer Corp.
Nylatron GS	Molysulfide nylon-6/6	Polymer Corp.
Nylon	Polyamides	E. I. duPont de Nemours & Co., Inc.
Nylon 200	Nylon	Firestone Tire & Rubber Co.
Nypel 1114MS	Molysulfide nylon-6/6	Nypel Inc.
Nysyn 30-5	Acrylonitrile–butadiene	Copolymer Rubber
Olefane	Polypropylene film	Avisun Corp.
Olemer	Propylene copolymer	Avisun Corp.
Olin CP-11 220	Nitrocellulose	Olin Corp.
Opalon	Polyvinyl chloride	Monsanto Chemical Co.
Oppanol	Polyisobutylene	Badische Anilin & Soda-Fabrik AG
Orthane 12036	Polyether based polyurethane elastomer-TPR	Ohio Rubber
Ortix	Poromeric film	Celanese Corp.
Oxiron	Epoxidized polybutadiene	Generic Name
Panarez	Hydrocarbon resins	Amoco Chemical Corp.
Panelyte	Laminates	Thiokol Chemicals Co.
Paracon	Polyester rubber	Bell Telephone Laboratories
Paracil BJLT-HX	Acrylonitrile–butadiene elastomer	Uniroyal, Inc.
Paracryl	Butadiene–acrylonitrile copolymer	U.S. Rubber Co.
Paradene	Coumarone-indene resins	Neville Chemical Co.
Paraplex P-19A:inject	Polyester	Rohm & Haas Co.
Parlon	Chlorinated rubber	Hercules Corp.
Parr 1D10	Diallyl-phthalate asbestos-reinforced	U.S. Prolam, Inc.
Parylene	Polyxylylene	Union Carbide Corp.
Parylene C	Poly p-xylylene	Union Carbide Corp.
Paxon AA55-003	Polyethylene	Allied Chemical Corp.
PDX-5229	20% glass-filled polyolefin elastomer TPR	LNP Corp.
PE-1007	Polyethylene	Gulf Oil Chemicals
Pearlon	Polyethylene film	Visking Corp.
Pee Vee Cee	Rigid polyvinyl chloride	ESB Corp.
Pelaspan	Expandable polystyrene	Dow Chemical Co.
PELLETHANE 2102-55D	Polyester polyurethane	Upjohn Co.
Penacolite G-1124	Resorcinol adhesive	Koppers Co.
Perbunan N	Butadiene–acrylonitrile copolymers	Farbenfabriken Bayer AG
Perspex	Acrylic resins	Imperial Chemical Industries, Ltd.
Petrothene	Polyethylene	National Distillers & Chemical Corp.

Table 14.2. (Continued)

Trade Name or Brand Name	Product	Manufacturer
Petrothene CR 90753	Polyolefin copolymer	U.S. Industrial Chem.
Pevalon	Polyvinyl alcohol	May and Baker Ltd.
Phenoweld 7	Phenolic adhesive	Hardman Inc.
Phenoxy	Polyhydroxy ether of bisphenol-A	Union Carbide Corp.
Philprene	Styrene–butadiene rubber	Phillips Petroleum Co.
PI-730	Glass-fiber-reinforced polyimide	Fiberite Corp.
Picco	Hydrocarbon resins	Hercules, Inc.
Piccocumaron	Hydrocarbon resins	Hercules, Inc.
Piccolyte	Terpene polymer resins	Hercules, Inc.
Plakon	Amino resins	Allied Chemical Corp.
Plaskon Alkyd AMC-446	Glass-flame-retardant alkyd	Allied Chemical Corp.
Plaskon DAP FA-10 V-0	Short-glass-flame-retardant diallyl-phthalate	Allied Chemical Corp.
Plaskon Epoxy MX-2342	Epoxy	Allied Chemical Corp.
Plaskon FR-1060	Polypropylene	Allied Chemical Corp.
Plaskon Urea FCG	Urea formaldehyde	Allied Chemical Corp.
Plastacele	Cellulose acetate flake	E. I. duPont de Nemours & Co., Inc.
Plastikon Putty No. 21	Natural rubber adhesive	B. F. Goodrich Chemical Co.
Plastilock 410	Polyurethane (TS) sealant	B. F. Goodrich Chemical Co.
Plastylene	Polyethylene	Pechiney-Saint-Gobain
Plenco	Phenolic resins	Plastics Engineering Co.
Plenco 118 Dk grey: Compr	Mineral-filled phenolic	Plastics Engineering Co.
Plenco 1500 Blk:Compr	Mineral alkyd	Plastics Engineering Co.
Plenco 713 Grey:Compr	Woodflour-filled melamine formaldehyde	Plastics Engineering Co.
Plexiglas	Acrylic sheets	Rohm & Haas Co.
Plexiglas	Acrylic	Rohm & Haas Co.
Pliobond H.T. 30	Nitrile adhesive	Goodyear Chemical Co.
Pliofilm	Rubber hydrochloride	Goodyear Tire & Rubber Co.
Plioflex	Polyvinyl chloride	Goodyear Tire & Rubber Co.
Plioflex PFR 1012	Styrene–butadiene elastomer	Goodyear Chemical Co.
Pliolite	Cyclized rubber	Goodyear Tire & Rubber Co.
Pliovic	Polyvinyl chloride	Goodyear Tire & Rubber Co.
Pluracol	Polyethers	Wyandotte Chemicals Corp.
Pluronic	Block polyether diols	Wyandotte Corp.
Plyfoam	PVC foam	
Plyophen	Phenolic resins	Reichhold Chemicals, Inc.
Polly Star HL 1.0	Polyethylene	St. Regis Paper Co.
Polyallomer	Ethylene block copolymers	Eastman Chemical Products
Polycarbafil F-50/20 faom	Short-glass-filled (20%) polycarbonate	Fiberfil/Dart
Polychem 100	Woodflour phenolic	Budd Chemical Co.
Polychem 201	Cotton-filled melamine formaldehyde	Budd Chemical Co.
Poly-eth	Polyethylene	Gulf Oil Corp.
Poly-eth Hi-D 9632	Polyethylene	Gulf Oil Corp.
Poly-eze	Ethylene copolymers	Gulf Oil Corp.
Polyfilm PSD 553	Polyethylene	Dow Chemical Co.
Polyflam RABS 678	ABS	A. Schulman Inc.
Poly-G 30-166	Polyether Polyol	Olin Corp.
Polylite	Polyester resins	Reichhold Chemicals, Inc.

Table 14.2. (Continued)

Trade Name or Brand Name	Product	Manufacturer
Polylite 31-382	Polyester	Reichhold Chemicals, Inc.
Polylumy	Polypropylene	Kohjin Co.
Polyman 507	Flame-retarded ABS/PVC	A. Schulman, Inc.
Polymin	Polyethyleneimine	Badische Anilin 6 Soda-Fabrick AG
Poly-pro	Polypropylene	Gulf Oil Corp.
Polyox	Water soluble resins	Union Carbide Corp.
POLYSAR Bromobutyl X2	Br-isobutene isoprene	POLYSAR
Polyseal BF-WO	Polyethylene	St. Regis Paper Co.
Polyset EPC 202	Cast/encap/potting	Morton Chemical
Polysystem 7544-02 Pt.A	Polyurethane foam-liquid	Olin Corp.
Polytrope TPP 220	Thermoplastic elastomer	A. Schulman Inc.
Polyvin 3502	Polyvinyl chloride	A. Schulman Inc.
Pool Paste	Epoxy adhesive	Atlas Mineral & Chemical
PPO	Polyphenylene oxide	Hercules, Inc.
Pro-fax	Polypropylene resins	Hercules, Inc.
Pro-fax PB-784	Polypropylene	Hercules, Inc.
Profil G-60/20	Long-glass polypropylene	Fiberfil/Dart
Propafilm C/110	Propropylene	Imperial Chemical Industries, Ltd.
Propathene	Polypropylene	Imperial Chemical Industries, Ltd.
Proxmelt A4063	Unspecified polymer hot melt	Pierce & Stevens
Pyralin	Cellulose nitrate	E. I. duPont de Nemours & Co., Inc.
Pyralin 3001 112	Glass-fabric-reinforced polyimide	E. I. duPont de Nemours & Co., Inc.
Pyrotex 110 RPD 9576	Asbestos-filled phenolic plastic	Raybestos Friction
R/flex 2411	Glass-polyester epoxy	Rogers Corp.
RAC 6250	Graphite-filled epoxy	Ciba-Geigy Comp.
Ravinil	Polyvinyl chloride	ANIC, S.P.A.
RCI 25000 Black	Woodflour-filled phenolic plastic	Reichhold Chemicals, Inc.
Reliabond RAC 7350	Glass-fabric-reinforced epoxy	Ciba-Geigy Comp.
Ren:C:O-Thane RP-6400	Poyurethane elastomer	REN Plastics
REN TWA-1040	Epoxy adhesive	REN Plastics
Resinite AF-50	Polyvinyl chloride	Borden, Inc.
Resinol	Polyolefins	Allied Resinous Products Inc.
Resinol Type A	Polyethylene	Norton Co.
Resinox	Phenolic resins	Monsanto Co.
Resistoflex	Polyvinyl alcohol	Resistoflex Corp.
Resiweld FE-004	Epoxy adhesive	H. B. Fuller Co.
Resloom	Melamine resins	Monsanto Co.
Restirolo	Polystyrene	Societa Italiana Resine
Resyn 30-9107	Polyester adhesive	National Starch
Revac AA 1030	Polyurethane foam-semifabric	Revertex, Ltd.
Rexene PE 102	Polyethylene	Rexene Polymers
Reynolon	Plastic films	Reynolds Metals Co.
Rezimac	Alkyds	Commercial Solvents Corp.
Rezl	Alkyd varnish	Sinclair-Koppers Co., Inc.
Rhoplex	Acrylic emulsions	Rohm & Haas Co.
Riblene	Polyethylene	ANIC, S.P.A.
Rigidex	Polyethylene	BP Chemicals (U.K.) Ltd.
Rigidite 5208	Carbon-fiber-filled epoxy	Narmco Materials
Rigolac	Polyester resins	Showa Highpolymer Co.
Rilsan	Nylon-11	Aquitaine-Organico
Rilsan AECN	Nylon 12	Rilsan Corp.

Table 14.2. (Continued)

Trade Name or Brand Name	Product	Manufacturer
RIMTHANE CPR2152-45D	Polyether polyurethane	Upjohn Co.
Rogers RX 1-501	Mineral-filled diallyl-phthalate	Rogers Corp.
Royal M5038	Natural rubber adhesive	Uniroyal, Inc.
Royalene 130	EPDM elastomer	Uniroyal, Inc.
Royalite	Thermoplastic sheet material	Uniroyal, Inc.
Royalite R12	ABS sheet	Uniroyal, Inc.
Roylar	Polyurethanes	Uniroyal, Inc.
Roylar A-855E2	Polyurethane elastomer	Uniroyal, Inc.
RTP 1003	Glass-fiber-reinforced polyester	Fiberite Corp.
Rucoblend 161	Polyvinyl chloride	Hooker Chemical & Plastic
Rucon	Polyvinyl chloride	Hooker Chemical Corp.
Rucothane	Polyurethanes	Hooker Chemical Corp.
Rulan	Flame-retardant plastic	E. I. duPont de Nemours & Co., Inc.
RX 1-540	Diallyl-phthalate	Rogers Corp.
Rynite 530	Polyethylene terephthalate	E. I. duPont de Nemours & Co., Inc.
Ryton R-10 Black 5002C	Polyphenyl sulfide	Phillips Petroleum
Saflex	Polyvinyl butyral	Monsanto Co.
Safom	Polyurethane foam	Monsanto Co.
Santolite	Sulfonamide resin	Monsanto Co.
Saran	Vinylidene chloride	Dow Chemical Co.
Saran 101	Polyvinylidene-chloride film	Dow Chemical Co.
Scotchbond	Unspecified polymer anaerobic adhesive	3M Co.
Scotchcast XR-5133	Filled epoxy	3M Co.
Scotchpak	Polyester film	3M Co.
Scotchpak 112	Polyester film	3M Co.
Scott Industrial Foam	Polyurethane	Scott Paper
Sealant U-5	Polyurethane (TS) sealant	Atlas Mineral & Chemical
Seilon	Thermoplastic sheets	Seiberling Rubber Co.
Selectron	Polyester resins	PPG Corp.
Selectron RS 50345	Polyester	PPG Industries
Shell DN 5028	Polypropylene	Shell Oil Co.
Silaprene M6324	Synthetic rubber sealant	Uniroyal, Inc.
Silastic	Silicone materials	Dow Corning Corp.
Silastic E RTV	Silicone elastomer-liquid	Dow Corning Corp.
Silastomer	Silicones	Midland Silicones Ltd.
Sirfen	Phenol-formaldehyde resins	Societa Italiana Resine
Sir-pel	Poromeric film	Georgia Bonded Fibers
Sirtene	Polyethylene	Societa Italiana Resine
Smooth-On EA-40	Epoxy adhesive	Smooth-On Inc.
Solithane LFTDI TR-90A	Polyurethane elastomer	Thiokol Chemical Co.
Solprene 1205	Styrene–butadiene	Phillips Chemical
Solvar	Polyvinyl acetate	Shawinigan Resins Corp.
Solvic	PVC	Solvay & Cie.
Somel 301G	Polyolefin elastomer	E. I. duPont de Nemours & Co., Inc.
Sonoco BB-1100	Polyvinyl alcohol adhesive	Sonoco Products Co.
Spandex	Polyurethane filaments	Generic Name (Formerly DuPont)
Spaulding Armite	Vulcanized fiber laminate	Spaulding Fibre
SPMC SP 323	Glass-filled polyester	Koppers Co.
S-polymers	Butadiene–styrene copolymers	Esso Labs
Spraythane	Urethane resin	Thiokol Chemical Corp.
Staflex	Vinyl plasticizers	Reichhold Chemical, Inc.
Standlite	Phenol-formaldehyde resin	Hitachi Chemical Co.

Table 14.2. (Continued)

Trade Name or Brand Name	Product	Manufacturer
Starex	Polyvinyl acetate	International Latex & Chemical Corp.
Stepanfoam A-210	Polyurethane foam-liquid	Stepan
Structo-Foam	Foamed polystyrene slab	Stauffer Chemical Co.
Strux	Cellular cellulose	Aircraft Specialties
Struxural-1	Polyurethane foam-semifabric	Strux Corp.
Stycast CPC-18	Polyurethane (TS)	Emerson & Cuming
Stymer	Styrene copolymer	Monsanto Co.
Styrafil	Fiberglass-reinforced polystyrene	Dart Industries, Inc.
Styrafil G 30/20	Long-glass-filled polystyrene	Fiberfil/Dart
Styraglas	Fiberglass-reinforced polystyrene	Dart Industries, Inc.
Styrex	Resin	Dow Chemical Co.
Styrocel	Polystyrene (expandable)	Styrene Products, Ltd.
Styrofoam	Extruded expanded polystyrene	Dow Chemical Co.
Styrofoam Insulation-B	Polystyrene foam-semifabric	Dow Chemical Co.
Styron 402D:Compr	Polystyrene	Dow Chemical Co.
Styron	Polystyrene	Dow Chemical Co.
Sulfil G-1500/10	Polysulfone	Fiberfil/Dart
Sullvac	Acrylonitrile–butadiene–styrene copolymer	O'Sullivan Rubber Corp.
Super Dylan	High-density polyethylene	O'Sullivan Rubber Corp.
Super Dylan FB-66	Polyethylene	ARCO/Polymer
Super-Lock 1	Methacrylate anaerobic	Devcon Corp.
Suprel 1220 UHMW	Polyethylene	Allied Chemical Corp.
Surlyn	Ionomer resins	E. I. duPont de Nemours & Co., Inc.
Surlyn 1554	Ionomer	E. I. duPont de Nemours & Co., Inc.
Swedcast	Acrylic sheet	Swedcast Corp.
Sylgard	Silicone casting resins	Dow Corning Corp.
Sylgard 182	Silicone cast/encap/potting	Dow Corning Corp.
Sylplast	Ureaformaldehyde resins	Sylvan Plastics, Inc.
Synpol E-BR 8407	Butadiene elastomer	Texas-U.S. Chemical Co.
Syntex	Alkyd resins	Celanese Corp.
Synthane	Laminated plastic products	Synthane Corp.
Synthane GEC 500	Glass-fabric-reinforced epoxy laminate	Synthane-Taylor
Taylorclad GEC 550	Glass-fabric-reinforced epoxy laminate	Synthane-Taylor
Tedlar	Polyvinyl fluorocarbon	E. I. duPont de Nemours & Co., Inc.
Tedlar 20 50 SG20TR	Polyvinyl fluoride film	E. I. duPont de Nemours & Co., Inc.
Teflon	Fluorocarbon resins	E. I. duPont de Nemours & Co., Inc.
Teflon A	Fluorocarbon film	E. I. duPont de Nemours & Co., Inc.
Teflon FEP	TFE copolymer	E. I. duPont de Nemours & Co., Inc.
Teflon TFE	Polytetrafluoroethylene	E. I. duPont de Nemours & Co., Inc.
Tefzel HT-2004	Glass-filled fluoropolymer	E. I. duPont de Nemours & Co., Inc.
Teglac	Alkyd coatings	American Cyanamid Co.
Tego	Phenolic resins	Rohm & Hass Co.
Telcar TR 100:Compr.	Polyolefin elastomer	B. F. Goodrich Chemical Co.
Tempreg	Low-pressure laminate	U.S. Plywood Corp.
Tenite	Cellulose derivatives	Eastman Kodak Co.
Tenite E8-354A	Polyethylene	Eastman Kodak Co.
Tenite 007 H	Cellulose acetate	Eastman Kodak Co.

Table 14.2. (Continued)

Trade Name or Brand Name	Product	Manufacturer
Tenite 155W	Polyethylene	Eastman Kodak Co.
Tenite 201 H	Cellulose butyrate	Eastman Kodak Co.
Tenite 306 H	Cellulose proprionate	Eastman Kodak Co.
Tenite 4E11	Propylene	Eastman Kodak Co.
Tenneco 2022	Polyvinyl chloride	Tenneco Chemical
Terluran	ABS polymers	Badische Anilin & Soda-Fabric Co.
Tetran	Tetrafluoroethylene	Pennsalt Chemical Corp.
Tetronic	Polyethers	Wyandotte Chemical Corp.
Texicote	Polyvinyl acetate	Scott Bader Co.
Texileather	Pyroxylin-leather cloth	General Tire & Rubber Co.
Texin	Urethane elastomer	Mobay Chemical Co.
Texin E-417	Polyurethane elastomer	Mobay Chemical Co.
Textolite	Laminated plastic	General Electric Co.
Textolite GEPEG	Glass-fiber epoxy laminate	General Electric Co.
TF 1001	Glass-fiber polyurethane elastomer TPR	LNP Corp.
TGDA-6765	Polystyrene	Gulf Oil Chemical
Thanol C-150	Polyether polyol	Jefferson Chemical Co.
Thermaflow	Reinforced polyesters	Atlas Powder Co.
Thermocomp	Glass-filled ABS	LNP Corp.
Thermogrip	Polyester hot melt	Bostik
Thiokol	Polyethylene sulfide	Thiokol Chemical Co.
Thixon 100	Synthetic rubber adhesive	Whittaker Corp.
Thurane	Polyurethane foam	Dow Chemical Co.
TMBD-2100	Polystyrene	Gulf Oil Chemical
Torlon 4000T	Polyamide-imide	Amoco Chemical Corp.
TPR-074	Thermoplastic elastomer copolymer	Uniroyal, Inc.
TPX	Poly(4-methylpentene-1)	Imperial Chemical Industries, Ltd.
Trans-4	*Trans*-1,4-polybutadiene	Phillips Petroleum Co.
Tremco Butyl sealant	Isobutylene isoprene	Tremco
Trevarno	Resin-impregnated cloth	Coast Mfg. & Supply Corp.
Trolen	Polyethylene	Dynamit Nobel AG
Trosiplast	Polyvinyl chloride	Dynamit Nobel AG
Tru-bond 203	Epoxy adhesive	Fenwal Inc.
Trucite 1000	Polystyrene film	Dow Chemical Co.
Trulon	Polyvinyl chloride resin	Olin Corp.
Tuffak	Polycarbonate	Rohm & Haas Co.
Tuf-flex 240	Rubber-modified polystyrene	American Hoechst
Tufmelt 3000	Unspecified polymer hot melt adhesive	Goodyear Chemical Co.
Tybrene	ABS polymers	Dow Chemical Co.
Tygon	Vinyl copolymer	U.S. Stoneware Co.
Tylose	Cellulose ethers	Farbwerke Hoechst AG
Tyril	Styrene–acrylonitrile copolymer	Dow Chemical Co.
Tyril 860 Compr.	SAN	Dow Chemical Co.
Ucon	Lubricants	Union Carbide Corp.
Udel	Plastic film	Union Carbide Corp.
Udel P-1700	Polysulfone	Ube Industries
Uformite	Urea resins	Rohm & Haas Co.
Ultrapas	Melamine-formaldehyde resins	Dynamit Nobel AG
Ultrathene	Finely divided polyolefins	National Distillers & Chemical Corp.
Ultrathene UE 630	Vinyl-acetate-filled polyethylene	U.S. Industrial Chemical

Table 14.2. (Continued)

Trade Name or Brand Name	Product	Manufacturer
Ultron	Vinyl film	Monsanto Co.
Ultryl	Polyvinyl chloride	Phillips Petroleum Co.
Union Carbide Carbowat 300	Glycol	Union Carbide Corp.
Unox	Epoxides	Union Carbide Corp.
Updown	Polychloroprene foam	
Upjon PAPI	PAPI	Upjohn Co.
Upolam PAPE	Nylon/polyethylene	Asko-Upo Oy
Urac	Urea-formaldehyde resins	American Cyanamid Co.
Urafil G-100/20	Fiberglass-polyurethane elastomer	Fiberfil/Dart
Uralene 1723-6-A/B	Polyurethane foam-liquid	Furane plastics
Uralane 5738 A/B	Polyurethane (TS) adhesive	M & T Chemical
Uralite 3111	Polyurethane elastomer	Hexcel Corp.
Urefoam R-02	Polyurethane foam-liquid	Atlas Mineral & Chemical
Uscolite	ABS copolymer	U.S. Rubber Co.
USP E-7148/G-10	Glass-fiber-filled epoxy	U.S. Polymeric
USS 200	Polystyrene	USS Chemicals
Uvex	Cellulose acetate butyrate	Eastman Kodak Co.
Valeron HS 3060	Polyethylene	Van Leer Plastics
Valox	Polybutylene terephthalate	General Electric Co.
Valox DR-455	Glass-fiber-filled polyester (PBT)	General Electric Co.
Vamac B-124 Masterbatch	Acrylic elastomer	E. I. duPont de Nemours & Co., Inc.
Vanguard 32-0005	Unspecified polymer adhesive	National Starch
Varcum	Phenolic resins	Reichhold Chemicals, Inc.
Varex	Polyester resins	McClosky Varnish Co.
Velan DF	Polyvinyl chloride	Firestone Tire & Rubber Co.
Velon	Polyvinyl chloride	Firestone Tire & Rubber Co.
Versamid	Polyamide resins	General Mills, Inc.
Vespel	Polymellitimide	E. I. duPont de Nemours & Co., Inc.
Vespel Parts KS-105 DF	Polyamide aromatic	E. I. duPont de Nemours & Co., Inc.
Vestamid	Nylon-12	Chemische Werke Huls AG
Vestolit	Polyvinyl chloride	Chemische Werke Huls AG
Vestyron	Polystyrene	Chemische Werke Huls AG
Vibrathane-B-600	Polyurethane elastomer	Uniroyal, Inc.
Vibrin	Polyester resins	Uniroyal, Inc.
Vinapas	Polyvinyl acetate	Wacker Chemie GmbH.
Vinoflex	PVC	BSAF Corp.
Vinol	Polyvinyl alcohol	Air Reduction Co.
Vinrez 6201	Acrylic adhesive	Celanese Polymers
Vinylfilm P1K	Polyvinyl chloride	Goodyear Tire & Rubber Co.
Vinylite	Polyvinyl chloride-co-vinyl acetate	Union Carbide Corp.
Vinyon	Polyvinyl chloride-co-acrylonitrile	Union Carbide Corp.
Vipla	Polyvinyl chloride	Montecatini Edison S.P.A.
Vistanex	Polyisobutylene	Enjay Chemical Co.
Vitel	Polyester resins	Goodyear Tire & Rubber Co.
Vithane	Polyurethanes	Goodyear Tire & Rubber Co.
Viton	Copolymer of vinylidene fluoride and hexafluoropropylene	E. I. duPont de Nemours & Co., Inc.
Viton A	Fluorocarbon elastomer	E. I. duPont de Nemours & Co., Inc.

Table 14.2. (Continued)

Trade Name or Brand Name	Product	Manufacturer
Voranol 3000	Polyether polyol	Dow Chemical Co.
Vorite 689M-2 System 27A	Polyurethane elastomer liquid	NL Industries
Vulcaprene	Polyurethane	Imperial Chemical Industries, Ltd.
Vulkollan	Urethane elastomer	Mobay Chemical Co.
Vybak	Polyvinyl chloride	Bakelite Xylonite, Ltd.
Vydyne	Nylon resins	Monsanto Co.
Vydyne R-108	Nylon-6/6	Monsanto Co.
Vygen 107	Polyvinyl chloride	General Tire & Rubber Co.
Vynathene VAE EY 903	Vinylacetate ethylene	U.S. Industrial Chemical
Vynex	Rigid vinyl sheeting	Nixon-Baldwin Chemicals, Inc.
Vyram	Rigid polyvinyl chloride	Monsanto Co.
Weatherban 101	Polysulfide sealant	3M Co.
Welvic	Polyvinyl chloride	Imperial Chemical Industries, Ltd.
Whirlclad	Plastic coatings	The Polymer Corp.
Whirlsint	Powdered polymers	The Polymer Corp.
Witcobond W-106	Polyurethane (TS) adhesive	Witco Chemical Corp.
XT Polymer	Acrylics	American Cyanamid Co.
Xylonite	Cellulose nitrate	B. X. Plastics, Ltd.
YF-1001	Glass-filled polyester elastomer TPR	LNP Corp.
Zendel	Polyethylene	Union Carbide Corp.
Zerok	Protective coatings	Atlas Minerals & Chemicals Div., ESB, Inc.
Zetafax	Polyethylene-co-acrylic acid	Dow Chemical Co.
Zetfin	Polyethylene-co-ethyl acrylate	Dow Chemical Co.
Zytel	Nylon	E. I. duPont de Nemours & Co., Inc.
Zytel ST-801	Nylon	E. I. duPont de Nemours & Co., Inc.
1,2-SBD XP-2625	Polybutadiene 1,2	Uniroyal, Inc.
3DER732/7DER331	Epoxy	Dow Chemical Co.
3M EC-1945 B/1	Epoxy primer	3M Co.
35DS	Polystyrene cast/encap/potting	Emerson & Cuming
5DER331/5DER542	Epoxy	Dow Chemical Co.
7DEN438/3DER599	Epoxy	Dow Chemical Co.

SUGGESTED READING

H. E. Atkinson, Ed., "Managing Corrosion Problems with Plastics," National Association of Corrosion Engineers, Houston, TX, 1977.

L. T. Britt, J. Pacitti, and J. R. Scott, *Chemical Resistance Data Sheets,* Rubber and Plastics Research Association of Great Britain, Vol. I and II, Shawbury, Shrewsbury, England, 1978.

R. F. Eisenberg, R. R. Kraybill, and R. B. Seymour, Chapter 1, "Selection of Materials for the Construction of Equipment," Techniques of Chemistry, E. S. Perry and A. Weissberger, Eds., Wiley-Interscience, New York, 1979.

O. H. Fenner, Chapter 4 in *Handbook of Plastics and Elastomers,* C. A. Harper, Ed., McGraw-Hill, New York, 1975.

S. Gross, Engineering Data Bank, *Modern Plastics Encyclopedia,* McGraw-Hill, New York, 1978.

M. E. Hammer, "Corrosion Data Survey," 5th ed., National Association of Corrosion Engineers, Houston, TX, 1976.

I. Mellon, *Corrosion Resistant Materials Handbook,* 3rd ed., Noyes Data Corp., Park Ridge, NJ, 1976.

National Association of Corrosion Engineers, "Coatings and Linings for Immersion Service," Katy, TX, 1972.

R. B. Seymour and R. H. Steiner, *Plastics for Corrosion Resistant Applications,* Reinhold Publishing Co., New York, 1955.

R. B. Seymour, "Chemical Resistance of Nonmetallic Materials," in Part 3, Vol. 3, *Treatise on Analytical Chemistry,* I. M. Kolthoff, P. J. Elving, and F. H. Stross, Eds., Wiley-Interscience, New York, 1970.

Index